# BIOLOGICAL INVESTIGATIONS

## LAB EXERCISES FOR GENERAL BIOLOGY

**Tenth Edition**

Edward Devine
*Moraine Valley Community College*

**KENDALL/HUNT PUBLISHING COMPANY**
4050 Westmark Drive    Dubuque, Iowa 52004

Copyright © 1973 by Virginia Smith, Calvin Kuehner, and Lenette Staudinger
Copyright © 1976 by Lenette Staudinger and Edward Devine
Copyright © 1979, 1981, 1983, 1993, 1994, 1996, 1998, 2001 by Kendall/Hunt Publishing Company

ISBN 0-7872-7203-5

All rights reserved. No part of this publication may be reproduced,
stored in a retrieval system, or transmitted, in any form or by any
means, electronic, mechanical, photocopying, recording, or otherwise,
without the prior written permission of the copyright owner.

Printed in the United States of America
10  9  8  7  6  5  4  3  2  1

# Contents

**Lab**

1. Methods of Science   1
2. Laboratory Measurements   5
3. The Making of Protein, Carbohydrates, and Lipids   15
4. Using the Microscope   27
5. Cell Structure: Plant and Animal Cells   33
6. Cell Diffusion   45
7. Factors Effecting Enzyme Activity   55
8. Effects of Environmental Variables on the Rate of Photosynthesis   59
9. Fermentation: A Jug of Wine   67
10. Mitosis and Meiosis   75
11. Investigating Heredity of Fruit Flies   87
12. Mendelian Inheritance Problem Solving   101
13. Effect of the Environment on Tobacco Plants   119
14. Human Genetic Traits   123
15. DNA—The Blueprint of Life   135
16. Genetic Engineering of Bacteria   143
17. The Lac Operon   149
18. Evidence of Evolution   153
19. Bacterial Resistance to Antibiotics   163
20. Hardy-Weinberg Calculations   169
21. Animal Classification   177

| | | |
|---|---|---|
| 22 | Electrocardiogram and Blood Pressure | 187 |
| 23 | Animal Structure: Dissection of the Fetal Pig | 197 |
| 24 | Feeding Activity in Small Animals | 217 |
| 25 | Kidney Function and Urinalysis | 223 |
| 26 | Chemical Effects on the Respiratory Rate of Brine Shrimp | 229 |
| 27 | Chemicals Effecting Muscle Contraction | 231 |
| 28 | Human Reflexes and Senses | 237 |
| 29 | Control Systems: Experiment in Temperature Regulation | 243 |
| 30 | Early Embryological Development | 251 |
| 31 | Effect of Hormone on Plant Growth | 255 |
| 32 | Chlorophyll Extraction and Separation | 261 |
| 33 | Diversity of Plants and Fungi | 267 |
| 34 | Population Ecology | 275 |
| 35 | Water Ecology Study | 287 |

Name _____  Section _____

# LAB 1

# Methods of Science

**Objectives**

   After completing this lab exercise the student will be able to:

1. List and define the steps of the scientific method.
2. State the purpose of an experiment.
3. Differentiate between experimental and control groups of an experiment.
4. Using a series of preliminary observations:
   a. state a problem developed from these observations
   b. formulate a hypothesis or probable solution to the problem
   c. design an experiment to test the hypothesis.

**Introduction**

   We live in a society that has changed drastically in the last 100 years. Many of these changes have been brought about by advances in science. Advances like the A-bomb, transistors, synthetic clothes and thousands of other items were developed in physics and chemistry. In the next 50-100 years the changes that will probably most drastically effect everyone's personal lives will be advancements in biology. These changes have already begun to take place with organ transplants, body control through biofeedback, environmental control, genetic engineering, cloned sheep, cloned cows, cloned pigs and thousands of other areas.

   A basic understanding of how science, specifically biology, functions will be a valuable assistance to citizens. Knowing the anatomy of a fetal pig enables us to understand the similarities between humans and other vertebrates. Knowing the evolutionary relationship of humans to other animals can allow a better understanding of how people fit into the environment. An understanding of how DNA operates would be useful in trying to decide if you want to support governmental control over genetic experiments. Knowing that ultraviolet light causes mutations would be helpful in deciding if controls should be placed on nitrogen oxide emissions from automobiles which destroys the ozone layer and increases our UV exposure.

   These problems and many others will be faced in the future. Can there be such a thing as too many people? At what point does this happen and what causes it? Once causes are established, be it for overpopulation or environmental deterioration, can probable solutions be devised and implemented? Answers to some of these questions will require a trained biologist. However answers to most of these questions and other problems will depend on the personal attitudes and understanding of everyone. Therefore, knowing how science functions is of importance. But more important is the realization that the technique a scientist employs in solving a problem is not limited to science. Many times this scientific technique or method is applied elsewhere.

Name _____    Section _____

## Part I: Scientific Method

Progress in science comes as a result of scientific investigations. The methods used in these investigations are many and varied. The scientific method is a vivid, dynamic process and cannot be reduced to a simple formula. The classification system used here is only a broad general view that is helpful in orientating a beginning scientist.

1. **Observation** of a phenomenon

    Something is noticed and attention is given to the observation. In this experiment, you will be given a sealed box. After an initial observation you will notice that there is something inside the box.

2. **Problem or Question**

    A problem is defined or a question is asked about the observation. Rather than simply accepting the observation and forgetting it, the questioning mind asks "why?" In this particular experiment, you might ask "what are the contents of the box?"

3. **Preliminary Information**

    In most types of scientific research, gathering preliminary information involves a literature search through books and journals. In this experiment, you gather preliminary information by tipping and shaking the box while listening to the sounds of the contents. As you listen you begin to get ideas about the contents of the box.

4. **Hypothesis**

    A hypothesis is a possible solution to the problem. It can also be explained as an "educated guess." After reviewing the preliminary information, a hypothesis is formed. The hypothesis must be stated in such a way that it is testable. To make a hypothesis statement like, "the contents of the box are pretty," is not adequate as it does not set up a basis of an experiment since it is not testable. A statement like "the box contains a round object" is more appropriate.

5. **Experiment**

    The primary purpose of the experiment is to test whether the hypothesis is correct. An experiment should test only one factor or variable while all other variables are kept constant. Every experiment must have a control or standard. A controlled experiment involves using two groups of the same kind of item or organism and treating them exactly the same except for the variable being tested. First test for the number of objects in the box. To get the most valid results there needs to be a comparison made to a constant standard. This is the control. In this case, the control would be an identical box with objects in it that you have set up. Conduct the experiment by comparing the sounds of the two boxes. Continued testing the contents of the box by considering such characteristics as number, size, shape, etc.

6. **Data**

    The results of data are collected from the experiment. Often the data are collected in a numerical form. Numerical data can be recorded in a table and plotted in a graph. Data can also be collected by visual and auditory means. These data may be recorded in the written notes of the researcher.

7. **Discussion**

    The data are analyzed and their meanings are interpreted. Comparisons are often made with the experiments and conclusions of other researchers.

8. **Conclusion**

    The conclusion summarizes the results of the experiment. The conclusion either accepts or rejects the hypothesis based on the data.

Name _____   Section _____

### TABLE 4.1
Microscope Magnifications

| Objective Lens | Ocular Lens | Total Magnificatioon | Diameter of the Field of Vision in | |
|---|---|---|---|---|
| | | | mm | Microns |
| Scanner 4X | 10X | 40X | 3.75 | 3,750 |
| Low-power 10X | 10X | 100X | 1.50 | 1,500 |
| High-poweer 43X | 10X | 430X | 0.35 | 350 |
| Oil immersion 100X | 10X | 1,000X | 0.15 | 150 |

With this information you can approximate the size of an object by estimating what fraction of the diameter of the field the object occupies. For example, if a spherical cell viewed under high power is about a third of the distance across the field of vision then the cell would be about 0.1 mm or 100 microns in diameter.

## Part IV: Cleaning the Compound Microscope

Cleaning the microscope lenses and slides before use is important for a clear, sharp image. Cleaning also prevents you from wasting time viewing dirt on the lens and slide.

1. **Only lens paper** should be used to clean the microscope lenses and prepared slides. Paper towels and facial tissue can scratch the lenses.
2. Carefully wipe the ocular and objective lenses with the lens paper.
3. Obtain a prepared microscope slide.
4. Using lens paper, clean the slide as you would clean sun glasses by applying light even pressure to both sides of the slide. Excessive pressure should never be used since the mounting medium never becomes completely hard.

## Part V: Focusing the Microscope

1. Using the coarse adjustment knob, raise the nosepiece until it stops.
2. Put the low power (10X) objective into the down or viewing position.
3. Examine the slide for the approximate position of the specimen by holding the slide up to the light.
4. Place the slide on the stage by moving the spring-loaded arm of the mechanical stage slightly to the side. DO NOT LIFT UP ON THE MECHANICAL STAGE ARM because it can be permanently bent. Release the spring-loaded arm against the side of the slide.
   *Note:* Do not put the arm over the slide, just against the side. Be sure the side of the slide closest to you is tight against the mechanical stage. This allows full view of the slide and eliminates slipping during fine movements.
5. Using the knobs on the mechanical stage, position the specimen under the objective.
6. Turn on the microscope light and open the iris diaphragm about one-fourth.
7. Using the coarse focus knob, position the objective all the way down.
8. While looking through the ocular, slowly turn the coarse focus up. When an object appears, focus in on it using the smaller fine focus knob. Adjust the light for the best image. If nothing appears, move to another spot on the slide and try again.
   *Note:* The image is inverted and reversed due to lens optics. Commercially prepared slides mount the specimens so they will be seen properly if there is a top/bottom or left/right orientation that is significant.
9. After the specimen is in good focus under low power, center it in the field of view.

Name _____    Section _____

10. Change to the high power (43X) objective by simply swinging that objective into the down or viewing position.

    *Note:* Good quality microscopes are **parcentered** and **parfocal**. **Parcentered** means the center of the field of vision stays essentially the same in each objective. **Parfocal** means practically no change in focus has to be made when changing objectives.

11. Adjust the fine focus and light intensity to obtain the best image.
12. Do not use the oil immersion (100X) objective at this time as special oil and focusing techniques are required.
13. Never force a gear or mechanism. If any part does not operate smoothly and easily, ask your instructor to check it.
14. The proper way to look through a microscope is with both eyes open. When you begin working on this technique, clear the lab table beside the microscope so you don't see any extraneous objects. Keeping both eyes open will be awkward at first, but you should adjust in 10 to 15 minutes of practice. When making drawings of specimens, you may find it useful to simultaneously look at the specimen with one eye and the drawing paper with the other eye. This technique will require a great deal of practice.

**Part VI: Drawing**

1. On the data sheet, identify and draw a small portion of the specimen that you have isolated on the slide.
2. When you have completed your microscope work, carefully:
    a. Remove the slide.
    b. Clean the lenses using lens paper.
    c. Put the lowest power objective in the down position.
    d. Position the mechanical stage so parts are not sticking out the side.
    e. Unplug the electrical cord and loop it in two or three large loops around the eye piece.
    f. Cover the microscope with the dust cover.
    g. Return the microscope to its properly numbered space in the microscope cabinet.
    h. Clean the lab table by wiping it with wet paper towels.

Name _____  Section _____

# LAB 18

# Evidence of Evolution

## Problem

Is there evidence that evolution has occurred?

## Objectives

After completing this lab exercise, the student will be able to:

1. Compare chick and pig embryos at similar stages of development.
2. Distinguish homologous and analogous structures.
3. Identify the bones in the front limbs of an amphibian, reptile, bird, and mammal.
4. Name the transition animal between reptiles and birds.
5. List reptile characteristics and bird characteristics shared by their transition animal.
6. Identify five bones in a vertebrate skull.
7. Note the changing position of the foremen magnum between vertebrate groups.
8. Arrange skulls of primates from primitive to modern.

## Preliminary Information

**Evolution** is the change in the allele (gene) frequency of a population over a period of time. The evolutionary patterns fall into two major categories: microevolution and macroevolution. In **microevolution**, the changes are minor. They might involve body coloration, wing length, bone length, enzyme production, or other similar traits. In **macroevolution**, the changes are major, resulting in substantially different body shape and functioning so a new species may result.

In both microevolution and macroevolution, when a new species arises, it will retain many of the characteristics of the population from which it arose.

There is evidence of evolution from many areas indicating that different types of organisms are related to one another. In most instances, relatedness does not necessarily imply a direct evolutionary line from one type to the next but rather an indication that the types present now are related through some common ancestral stock.

This laboratory exercise will investigate some lines of evidence for evolution.

## Part I: Evidence from Embryology

If organisms are related, their embryological development should be similar.

1. Obtain prepared microscope slides of:
   a. pig embryo
   b. chick embryo.
   These embryos are at comparable stages of development.

Name _____ Section _____

2. Using a **binocular microscope**, rotate the slide until the head is up, examine and diagram each embryo.
3. Label these parts on both:

   **head** (has 2 or 3 brain bulges)
   **eye** (dark, round)
   **front limbs** (small bulge near the heart)
   **hind limbs** (small bulge near the tail)
   **heart** (a prominent bulge near the middle)
   **spinal cord**
   **tail**

Vertebrate Embryos

| Chick Embryo | Pig Embryo |
|---|---|
|  |  |

## Part II: Evidence from Limb Structure

If organisms are related, their limbs should have similar bones. **Homologous structures** have similar shapes, development, and relationships with other surrounding structures. The bones in the limbs of vertebrates are homologous structures. Analogous structures perform similar functions. A bird wing and a butterfly wing are analogous structures. Homologous structures indicate relationships between organisms while analogous structures do not.

1. Observe and diagram the forelimbs of the frog, turtle or alligator, chicken, bat, cat or goat, and human.
2. Label these bones on each diagram:

   **humerus** (first bone in the upper arm)
   **radius** (forearm bone on the thumb side)
   **ulna** (other bone in the forearm)
   **carpals** (wrist bones)
   **metacarpals** (long bones in the hand)
   **phalanges** (three finger bones).

Name _____   Section _____

## Vetebrate Forelimbs

| Frog | Turtle or Alligator |
|---|---|
| Chicken | Bat |
| Cat or Goat | Human |

Name _____    Section _____

## Part III: Evidence of Evolution from a Transition Animal

A transition animal provides a line of relatedness.

The punctuated equilibrium theory of evolution states that the change from one species to another occurs very quickly in geological terms. The change happens so quickly that the transition forms are usually not found in the fossil record. The transition forms may not have been fossilized or the fossil remains may not have been discovered. Transition forms would certainly provide good evidence of evolution. *Archaeopteryx lithographica* is a transition animal between reptiles and birds.

1. Observe the fossil of *Archaeopteryx lithographica* at the demonstration area.
2. Complete the table below by listing some of the reptile characteristics and bird characteristics that this organism possesses.

Characteristics of *Archaeopteryx*

| Reptile Characteristics | Bird Characteristics |
|---|---|
|  |  |
|  |  |
|  |  |
|  |  |

## Part IV: Evidence from Vertebrate Skull Structure

Similarities and differences in skulls show relatedness yet adaptation to an upright posture.

1. Observe the vertebrate skulls at the demonstration area.
2. Locate the following parts in each skull:

   **maxilla** (major upper jaw bone, holds back teeth)
   **mandible** (lower jaw)
   **frontal** (major bone over the eye)
   **parietal** (major bone over the brain)
   frontal and parietal are fused in frog forming frontoparietal
   **foramen magnum** (hole for spinal cord).

Name _____  Section _____

3. Note the position of the **foramen magnum** on each skull by diagramming each skull from a side view and indicating the position of the foramen magnum with an arrow.

## Position of Foramen Magnum

| Turtle | Bird |
|---|---|
| | |

| Dog or Fox | Chimp or Gorilla |
|---|---|
| | |

| Australopithecus | Human |
|---|---|
| | |

Name _____  Section _____

4. Complete the table below to correlate some aspects of skull anatomy with posture. For each skull, note:
   a. eye orientation as:
      sideways
      forward
   b. position of foramen magnum as:
      behind
      angular
      below
   c. posture as:
      4-legged
      2-legged, stooped
      2-legged, upright.

**TABLE 18.1**
Some Aspects of Skull and Body

| Animal | Eye Orientation | Foramen magnum | Posture |
|---|---|---|---|
|  |  |  |  |
|  |  |  |  |
|  |  |  |  |
|  |  |  |  |
|  |  |  |  |
|  |  |  |  |

## Part V: Evidence from Primate Skull Structure

This portion of the lab may be done as a class activity.

Anthropologists use many techniques in examining skulls to help them determine the degree of relatedness between animals. Some of these techniques involve complex mathematical measurements and comparisons involving mathematical data. They also use overall visual comparisons in their examinations.

In this portion of the lab, visual comparisons will be used to place the skulls in a line from primitive to modern. This does not necessarily imply that one group evolved from another but it can be used, along with other data, to show a degree of relatedness.

1. Examine the skulls of the monkeys, apes, and humans at the demonstration area.
2. Make a list of characteristics that can be used to distinguish and separate the skulls from one another.

Name _____    Section _____

3. Some Characteristics Used to Distinguish Skulls

   a.

   b.

   c.

   d.

   e.

4. Arrange the skulls in a linear order from primitive to modern. All members of the class do not have to agree on the order.
5. Name the skulls in the order you have placed them. Name primitive first, end with modern. The names of the skulls are on the back of the skulls.

**Arrangement of Primate Skulls**

Name _____  Section _____

**Part VI: Discussion**

1. The embryos of all vertebrates go through many similar stages of development. What does this suggest?

2. You do not have a tail. Would you expect a tail to be present in the human embryo if you examined one at the same stage of development as the chick and pig? Explain.

3. Distinguish between homologous and analogous structures. Which type indicates a relationship between organisms?

4. As you examined the bones in the forelimbs of the various animals, what bones in the forelimbs changed:
    a. the most?

    b. the least?

5. Select one animal you examined and explain how the shape of the bones in its forelimb is related to the type of movement or use.

6. Why is *Archaeopteryx* considered a good transition animal?

Name _____   Section _____

7. In an animal with an upright posture, what is the advantage of having the foramen magnum toward the middle of the skull rather than at the back of the skull?

8. As the overall picture of skull development from monkey to human is considered, discuss trends that appear regarding:

   a. shape of the face

   b. canine tooth structure

   c. bony projections on the skull

   d. strength of the lower jaw

   e. balance of the skull on the vertebral column.

Name _____    Section _____

# LAB 19

# Bacterial Resistance to Antibiotics

## Problem

Can bacteria become resistant to antibiotics?

## Objectives

After completing this lab exercise, the student will be able to:

1. Demonstrate sterile technique when streaking and handling a bacterial culture plate.
2. Determine bacterial resistance to antibiotics.
3. Demonstrate the proper incubation procedure for a bacterial culture plate.
4. Name an environmental factor that can induce a mutation.
5. Discuss the relationship between mutations, environmental selection pressures, and evolution.

## Preliminary Information

**Evolution is a change in the gene frequency of a population over a period of time**. To put it another way, evolution is the change in the percentage of one gene compared to its alternate in a population over many generations. **If the percentage of any characteristic in a population changes over time then the population has evolved**. Some changes in a population may go almost undetected while other changes may be quite dramatic. Each individual in a population has a set of genes which enables it to function in its particular environment. A mutation is an alteration in functioning genetic instructions. In most instances, this alteration will cause a detrimental effect when the genetic instructions stop functioning properly. However, in a very small percentage of mutations, the genetic change may be advantageous, making the individual better adapted to the environment, especially if the environment is changing in such a way that it selects for the mutation.

It is important to understand that:

1. an individual may undergo a mutation but it is the population that evolves.
2. the environment always selects for preexisting characteristics in the population. The selection factor can not cause the mutation.
3. a mutation is most often detrimental.

Bacteria are often used in experimentation because large populations can be grown in a small space, and a new generation can be produced every 20 minutes under optimum conditions. Bacteria do not need to mate to reproduce but simply divide in half by a process called binary fission.

In this experiment, the resistance of a population of bacteria to antibiotics will be tested. The preexisting resistance along with any resistance which may coincidentally occur by ultraviolet light induced mutations will be tested.

Name _____    Section _____

## Part I: Methods and Materials

### Materials

The following will be needed for each team of four or as the instructor directs:

    4 nutrient agar plates
    1 sterile cotton swab
    1 antibiotic disperser
    1 permanent marker
    1 ultraviolet light
    1 Bunsen burner
    1 forceps
    1 small beaker of sterilizing solution
    1 liquid culture of bacteria

### Procedure

1. Wash and dry the top of the lab table. Keep it clear of all personal items.
2. Obtain 4 nutrient agar plates for each lab table. Keep the lid closed.
3. Using a permanent marker and small printing, label the outer edge on the agar side of the plate with your name, date, and experiment.

**Note: Sterile Procedure and Streaking**

- Do not remove the lid completely from the petri plate. To streak with bacteria, just lift the edge and insert the swab.
- Streak back and forth covering the entire plate then turn the plate 90 degrees and repeat without rewetting the swab. This ensures that the entire surface of the agar is evenly covered with bacteria.

**Note: Wear safety glasses.**

4. Obtain a liquid culture tube of bacteria.
5. Light the Bunsen burner. Carefully remove the cap from the culture tube and flame the top of the tube to kill any stray bacteria by passing it through the flame of the Bunsen burner. Keep the cap in your hand.
6. Dip the sterile swab into the liquid bacterial culture, stir the culture, press the swab against the inside of the tube to get the swab as dry as possible.
7. Flame the top of the tube again, replace the cap, and pass the tube on to the next table.
8. Lift the top of the petri plate just enough to insert the swab.
9. With the swab, spread the bacteria evenly over the agar plate. Rotate the plate 90 degrees (a quarter turn) and repeat streaking. Streak all of the petri plates in this fashion.
10. **Immediately** put the swab into the sterilizing solution.

Name _____  Section _____

11. Each bacterial plate will be exposed to ultraviolet light for a different amount of time. The exposure times are 15, 30, and 60 seconds. Write the exposure time on the bottom edge of each plate.
12. To expose the plates to ultraviolet light:
    a. place the plate on the table,
    b. remove the lid, hold it up-side-down,
    c. place the ultraviolet light over the petri plate,

    *Caution.* Ultraviolet light can damage the eyes. Never look into the light.

    d. turn the ultraviolet light on for the required amount of time,
    e. place the lid back on the petri plate.
13. The instructor will dispense the antibiotics by placing the antibiotic disc dispenser over the opened petri plate and pressing the handle down firmly to dispense the antibiotic discs. Each disc has the antibiotic name and concentration printed on it.
14. Quickly cover the petri plate.
15. Let the plates sit right-side-up for 10 minutes so the discs make firm contact with the agar.
16. Place the bacterial plates, up-side-down in the tray for your class in the incubator at 37 degrees Celsius for two days.
    Note: Plates must be cultured and stored up-side-down so water does not condense on the bacteria side of the plate.
17. Wash and dry the lab table.
18. Wash your hands.

## Part II: Data

1. After incubating, place the petri plates on the table in order of ultraviolet exposure time then turn them so each antibiotic disc is in the same relative position so easy comparisons can be made.
2. Record the antibiotics and concentrations you used in the data table.
3. Examine the plates and record whether the bacteria are:
    sensitive (clear zone around antibiotic)
    intermediate (halo or spotty colonies in clear zone)
    resistant (growth up to antibiotic).
4. In the table, also record (in parentheses) the number of colonies found in the clear areas surrounding any antibiotic disc.
5. Remember the antibiotic did not cause the mutation, it only selected for those bacteria which already possessed the mutation for resistance to a specific antibiotic.

**TABLE 19.1**

Antibiotic Disc Symbols

| Symbol | Antibiotic | Concentration |
|---|---|---|
| AM | Ampicillin | 10 ug |
| C | Choloramphenicol | 30 ug |
| E | Erythromycin | 15 ug |
| K | Kanamysin | 30 ug |
| N | Neomycin | 30 ug |
| P | Penicillin | 2 units |

Name _____ Section _____

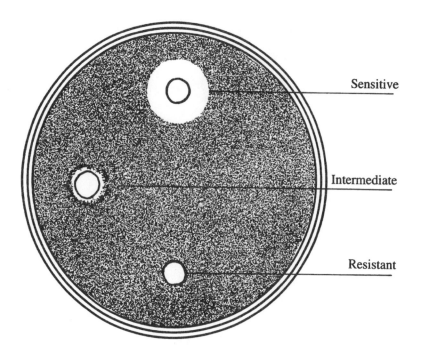

**Figure 19.1** Bacterial Growth with Antibiotic

**TABLE 19.2**
Antibiotic Effectiveness

| Antibiotic/ Concentration | Ultraviolet Exposure Time | | | |
|---|---|---|---|---|
| | 0 seconds | 15 seconds | 30 seconds | 60 seconds |
| | | | | |
| | | | | |
| | | | | |
| | | | | |
| | | | | |
| | | | | |
| | | | | |
| | | | | |
| | | | | |
| | | | | |

Name _____  Section _____

## Part III: Discussion

1. Why is it important to keep the lid on the nutrient agar plate except when direct access to the plate is required?

2. When streaking an agar plate, why streak one direction then turn the plate 90 degrees and repeat again?

3. List the antibiotics from most to least effective on this strain of bacteria.

4. Name the antibiotics to which the bacteria are:
   a. completely resistant.

   b. partially resistant.

5. If antibiotic resistance appears in bacteria, did the antibiotic cause the mutation? Explain.

6. What effect does increased exposure to ultraviolet light have on bacterial growth? Why do you think this happens?

Name _____  Section _____

# LAB 20

# Hardy-Weinberg Calculations

**Problem**

Can the percentages of dominant and recessive genes in a population be determined?

**Objectives**

After completing this lab exercise, the student will be able to:

1. Explain the significance of the Hardy-Weinberg Equation.
2. State the conditions that must be present for the allele frequencies to remain constant from generation to generation.
3. Calculate the percent of dominant and recessive genes in a population from the phenotypic ratios.
4. Predict the phenotypic and genotypic ratios in successive generations when the starting generation is known.

**Preliminary Information**

In the early 1900s, the field of population genetics began to be studied. In 1908, G. H. Hardy, an English mathematician, and G. Weinberg, a German physician, developed independently and published almost simultaneously a mathematical explanation of why certain genetic traits persist in a population generation after generation. Their work shows that dominant genes do not replace recessive genes. Their mathematical expression is called the Hardy-Weinberg Equation. According to the **Hardy-Weinberg Equation**, the percentages of dominant and recessive genes in a population will remain the same generation after generation if these conditions are met:

1. Large population
2. Isolated population
3. Random mating
4. No mutation
5. No natural selection.

If these conditions are met, then the percent of genes in a population cannot change and evolution cannot occur. Evolution is the change in the allele frequencies of a population over a period of time. However, if the percentages of genes are not the same generation after generation, then evolution has occurred.

The Hardy-Weinberg Equation can be used when there are only two genes involved for one trait. One example of the use of the Hardy-Weinberg Equation can be seen in a fruit fly population. The equation can begin with the percent of each type of fruit fly present in the experimental population. For example, if a population is set up with 7 homozygous dominant wild-type (gray body) flies and 3 homozygous recessive

Name _____  Section _____

ebony-body flies, then 70 percent of the population has gray bodies and 30 percent of the population has ebony bodies. This can be expressed mathematically like this:

**TABLE 20.1**
Proportion to Percent

| Trait | Proportion | Percent | Decimal Equivalent |
|---|---|---|---|
| Wild | 7/10 | 70% | .7 |
| Ebony | 3/10 | 30% | .3 |

The Decimal Equivalent is Used in Hardy-Weinberg Calculations.

If these flies breed randomly, the next generation can be calculated by either of two methods: the Punnett square method or the binomial expansion method. In either of these methods "p" represents the frequency of dominant genes in the population for a particular trait and "q" represents the frequency of recessive genes in the population for the same trait. Since there are only two genes for a trait, then:

$p + q = 1.$

In other words, the percentage of dominant genes for a trait plus the percentage of recessive genes for the same trait in a population equal 100 percent of the possible genes for the trait.

## Punnett Square Method

|   | p | q |
|---|---|---|
| p | pp | pq |
| q | pq | qq |

now add decimal equivalents

|   | .7 | .3 |
|---|---|---|
| .7 | .49 | .21 |
| .3 | .21 | .09 |

## Binomial Expansion Method

```
    p + q
  X p + q
   pp + pq
        pq + qq
   pp + 2pq + qq
```

now add decimal equivalents

```
     .7 + .3
  X  .7 + .3
    .49 + .21
          .21 + .09
    .49 + .42 + .09
```

or   $p^2 + 2pq + q^2$

The interpretation of either the Punnett square method or the binomial expansion method is the same:

- 49% of the population is homozygous dominant
- 42% of the population is heterozygous
- 9% of the population is homozygous recessive.

## Examining a Population

If the genes for a characteristic are dominant-recessive, there is no way to distinguish homozygous dominant and heterozygous. Therefore, **the beginning point in examining the allele frequency of a population is the homozygous recessive.** When the percent of homozygous recessive is known, it is possible to work backward to calculate the percent of recessive genes in the population, and from there the percent of dominant genes.

Name _____  Section _____

In the Hardy-Weinberg Equation, the **percent of genotypes** in the population is represented by the equation:

$p^2 + 2pq + q^2 = 1$ or
Homozygous + Heterozygous + Homozygous = 100% of the possibilities for the trait
   dominant                                   recessive

The **percent of genes** in the population is represented by the equation:

$p + q = 1$ or
Dominant genes + recessive genes = 100% of the genes in the population for the trait

Now examine the fruit fly numbers using the Hardy-Weinberg Equation. It says that 9 percent of the population ($q^2$) will show the recessive trait. However, since many of the recessive genes are present in the heterozygotes, the total percentage of recessive genes (q) in the population can be calculated by taking the square root of $q^2$. In this case:

if $q^2 = .09$
then $q = .3$

which means that 30 percent of the genes in this population are recessive.

Once the frequency of the recessive gene has been calculated, the frequency of the dominant gene can be calculated by the formula:

$p = 1 - q$

In this case,

$p = 1 - .3$
$p = .7$

Name _____  Section _____

Once the gene frequencies are know, the Punnett square can be completed.

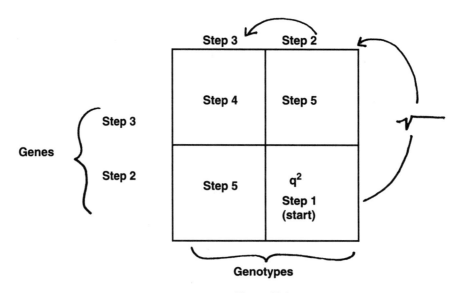

**Figure 20.1**
Steps to Examining a Population

**TABLE 20.2**
Hardy-Weinberg Summary

| Frequency of Genotypes | $p^2 + 2pq + q^2 = 1$ |
|---|---|
| Frequency of Genes | $p + q = 1$ |

## Part I: Examination of a Human Population

### Right-Left Handedness

Humans possess a few characteristics with a dominant-recessive relationship. In this portion of the lab, the class will be the population.

1. How many students are in the class? _____
2. In humans, right handedness is dominant over left handedness. How many left handed people are in the class? _____
3. What percent of the class is left handed? _____

   This percent, in decimal form, is $q^2$. Write it here _____ and in the Punnet square.

4. Calculate q by taking the square root of $q^2$. Put q on the Punnet square.

Name _____  Section _____

5. Calculate p (p = 1 - q). Put p on the Punnett square.

6. Fill in the Punnett square.

7. Summarize the data:

**TABLE 20.3**
Frequency of Right- and Left-Handedness

| $p^2$ homozygous Right | 2 pq Heterozygous Right | $q^2$ Homozygous Left |
|---|---|---|
|  |  |  |
| p % Dominant Genes | q % Recessive Genes |  |
|  |  |  |

### Earlobes

1. How many students are in the class? _____
2. In humans, free earlobes are dominant over attached. How many people have attached earlobes? _____

3. What percent of the class has attached earlobes? _____

   This percent, in decimal form, is $q^2$. Write it here _____ and in the Punnett square.

4. Calculate q by taking the square root of $q^2$. Put q on the Punnett square.

5. Calculate p (p = 1 - q). Put p on the Punnett square.

6. Fill in the Punnett square.

7. Summarize the data:

**TABLE 20.4**
Frequency of Earlobes

| $p^2$ homozygous Free | 2 pq Heterozygous Free | $q^2$ Homozygous attached |
|---|---|---|
|  |  |  |
| p % Dominant Genes | q % Recessive Genes |  |
|  |  |  |

Name _____ Section _____

## Hairline

1. How many students are in the class? _____
2. In humans, widow's peak is dominant over straight hairline across the forehead. How many people have straight hairline? _____

3. What percent of the class has straight hairline? _____

   This percent, in decimal form, is $q^2$. Write it here _____ and in the Punnett square.

4. Calculate q by taking the square root of $q^2$. Put q on the Punnett square.

5. Calculate p (p = 1 - q). Put p on the Punnett square.

6. Fill in the Punnett square.

7. Summarize the data:

**TABLE 20.5**
Frequency of Hairlines

| $p^2$ homozygous Widow Peak | 2 pq Heterozygous Widow Peak | $q^2$ Homozygous Straight |
|---|---|---|
|  |  |  |
| **p % Dominant Genes** | **q % Recessive Genes** |  |
|  |  |  |

## Part II: Predicting Later Generations

1. Select one of the traits examined in Part I. What trait do you want to use here? _____
2. The members of this class will be the parent generation.
3. Use the data from Part I. What **percent of the genes** in this class population are:

   dominant? _____

   recessive? _____

4. Hypothetically mate the members of this class to produce the $F_1$ by filling in the Punnett square with decimal equivalents of the percentages from above:

Name _____ Section _____

5. In the F₁ generation, what percent are:

    homozygous dominant? _____

    heterozygous? _____

    homozygous recessive? _____

6. In the F₁ population, what is the frequency of:

    dominant genes (p)? _____

    recessive genes (q)? _____

7. Hypothetically mate the members of the F₁ generation to obtain the F₂ by filling in the Punnett square:

|   | p ____ | q ____ |
|---|--------|--------|
| p ____ |    |    |
| q ____ |    |    |

8. In the F₂ generation, what percent are:

    homozygous dominant? _____

    heterozygous? _____

    homozygous recessive? _____

9. In the F₂ population, what is the frequency of:

    dominant genes (p)? _____

    recessive genes (q)? _____

10. Does the gene frequency change in this population from generation to generation? Discuss.

Name _____  Section _____

## Part III: Discussion

1. What conditions must be present for the percentages of genes in a population to remain constant from generation to generation?

2. As you examine a population over several generations, how can you tell if a population is evolving?

3. When you examine a population using the Hardy-Weinberg Equation, why must you begin with the percentage of individuals with the recessive trait rather than those with the dominant trait?

4. A recessive trait is seen in 4% of a population. Using this as a starting point, complete the Punnett square and the table below. Then answer the associated questions.

**TABLE 20.6**
Hardy-Weinberg Population Analysis

| p | q | $p^2$ | 2pq | $q^2$ |
|---|---|---|---|---|
|   |   |   |   |   |

a. What percent of the population is homozygous dominant? _____

b. What percent of the population is heterozygous? _____

c. What percent of the genes in the population are dominant? _____

d. What percent of genes in the population are recessive? _____

Name _____  Section _____

# LAB 21

# Animal Classification

## Problem

How can organisms be grouped into systematic categories according to their physical characteristics?

## Objectives

After completing this laboratory exercise, the student will be able to:

1. List in proper order, the taxonomic sequence from kingdom through species.
2. Determine if an organism has radial, bilateral or no symmetry.
3. Determine if an organism has an endoskeleton, an exoskeleton or no skeleton.
4. Recognize complete or partial segmentation.
5. Identify an organism into kingdom, phylum and class according to the taxonomic key provided in this lab.

## Preliminary Information

One of the first persons to group organisms was Aristotle. Aristotle grouped animals according to where they lived. This method was confusing and did not show relationships. For example, a whale, a shark, and an octopus would be classified together. But it was not until the 1700s that *taxonomy* or the science of classification really developed. During this time, a Swedish botanist, Carolus Linneaus, developed the system of *binomial nomenclature* or two names. The advantage of this system is that every living thing is given only one name and no two kinds of organisms can have the same name. This eliminates the confusion of common names. For instance, the bird Americans refer to as a robin is called a thrush in England, and in Europe it has a different name in every country. However, if you use the name *Turdus migratorius,* it can mean only one type of bird, no matter if you are American, English, Russian, or Japanese.

## Part I: Taxonomic Rank

Placing organisms in categories can be like sorting mail.

    Kingdom                    Country
      Phylum                   State
        Class                   Town
          Order              Street number
            Family          House number
             Genus          Last name
               Species          First name

Name _____    Section _____

|  | **Human** | **Wolf** | **Herring Gull** | **Red Oak** |
|---|---|---|---|---|
| Kingdom | Animalia | Animalia | Animalia | Plantae |
| Phylum | Chordata | Chordata | Chordata | Magnoliophyta |
| Class | Mammalia | Mammalia | Aves | Angiospermae |
| Order | Primates | Carnivora | Charadriiformes | Fagales |
| Family | Hominidae | Canidae | Laridae | Fagaceae |
| Genus | *Homo* | *Canis* | *Larus* | *Quercus* |
| Species | *sapiens* | *lupus* | *argentatus* | *ruba* |

The importance of this type of a classification scheme is that it shows the relationship between organisms So if two organisms belong to the same family, but different genera, they are more closely related than two organisms that belong to the same order, but different families.

1. Of the other three organisms classified in the above table, which is most closely related to humans? Why?

2. Of the other three organisms classified in the above table, which is least related to humans? Why?

## Part II: Recognition of Characteristics

Before you begin using a key, you should be able to recognize certain characteristics. Use the demonstration specimens and animal set specimens. Complete the following using the numbers on the jars for identification or match number with name by using the list at the end of this lab.

**Single-celled organisms or colonies**: Single-celled organisms exist by themselves. Colonies are groups of single-celled organisms attached and living together but physiologically independent. Each cell in a colony looks like every other cell. The cells are not specialized into different functions. Colonies often form thread-like strands. Go to the demonstration table, examine the specimens and name the ones that are single-celled or colonies. _____

**Radial symmetry**: Many lines could be drawn top to bottom through the middle of an animal and two identical halves would result. These organisms tend to be circular or radiate out from a central point. Name three organisms with radial symmetry. _____

**Bilateral symmetry**: Only one line can be drawn top to bottom through the middle of an organism that will result in two identical external halves. For instance, in humans the only way you can get two identical halves is by passing a line down the middle through the nose, mouth, etc. Name three other organisms with bilateral symmetry. _____

Name _____    Section _____

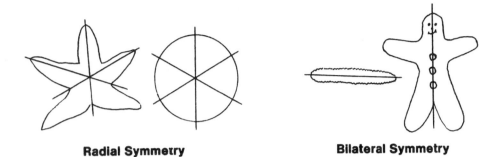

**Radial Symmetry**     **Bilateral Symmetry**

**Figure 21.1** Symmetry

**No symmetry**: No particular shape. Name one organism that lacks symmetry. _____

**Segmentation**: Segments are repeating units that are similar. With complete segmentation the units are obvious, like connected railroad cars. Incomplete segmentation is not so obvious. The segmentation need not be external, it may be part of the skeleton such as the backbones in a human.

Name three organisms that show some degree of segmentation, internal or external. _____
_____

**No Segmentation**: Name three organisms that lack any visible segmentation. _____
_____

**Exoskeleton**: Hard outer covering, these animals will usually go crunch when stepped on. Name three.
_____

**Endoskeleton**: Internal bony structure. Examine demonstrations of animal skeletons. Name three animals that have an endoskeleton. _____
_____

Name _____  Section _____

**Part III. Grouping of Organisms into Kingdoms**

The diversity of living things is far too great for anyone to know them all, but the differences are not endless—patterns are discernible. Organisms generally have several obvious characteristics which enable us to differentiate one from another. For example, it is not difficult to distinguish among a clam, a lobster and a mouse. It is obvious that they have more in common with each other than they have in common with an oak tree or a bacterium.

When organisms are grouped into related groups, it is best to begin with the most basic common characteristic and then proceed to the more specific but sometimes obscure details.

Go to the demonstration table to observe the following organisms. Classify them into the appropriate kingdom using the following key.

|  | Kingdom |  | Kingdom |
|---|---|---|---|
| bacteria | _____ |  |  |
| Spirogyra | _____ | green plant | _____ |
| Oscillatoria | _____ | mushroom | _____ |
| Paramecium | _____ | planaria | _____ |

**How to Use a Key**

Most keys begin with some obvious characteristic and present you with at least two choices. In the kingdom key you must first decide if the organism is single or multicellular. A single cell is so small that you need a microscope to see it whereas multicellular organisms can be seen without a microscope. Do not confuse a colonial organism made of clusters or a chain of similar cells with a multicellular organism. If it is single-celled you are referred to item number 2 for the next two choices. Select the item that best fits the organism in question. However, if the organism is multicellular you are referred to item number 3 where you must decide if it produces its own food or not.

**Key to Kingdoms**

1a. Single-celled organisms or colonies of similar cells; usually microscopic ..................... 2
1b. Multicellular organisms; body composed of tissues or layer of cells; usually macroscopic ......... 3

2a. Cells with a nucleus................................................. **Kingdom Protista**
2b. Cells without nucleus............................................... **Kingdom Monera**

3a. Produces own food, usually green; lacks self-mobility ..................... **Kingdom Plantae**
3b. Does not produce own food................................................. 4

4a. Plant-like in appearance, never green, feeds on decaying matter................ **Kingdom Fungi**
4b. Mobile, eats other organisms........................................... **Kingdom Animalia**

Name _____   Section _____

## Part IV: Classification of the Animal Kingdom

Examine all of the specimens in the set. Notice they are all animals. So now you are ready to proceed to the next level of classification, the phylum.

The first two phyla groups have been suggested for you. Separate all the jars into separate groups that fit this description. Write the number of the jar after the appropriate description. These two phyla will consist of *more than half* of the jars. After you have completed these two groups to your satisfaction, have the *instructor check* before proceeding.

Now work with the remaining jars. Place those that look alike together. Devise a description of the group and record it as phylum number 3. Again record the number of the jars. These remaining groups will not be as large as the first two. Make sure all organisms are classified. It is important that you describe how the organism looks. For example, "long, slender, round, with segments" is fine. But it is not permissible to state "earthworm" as it does not tell you how an earthworm looks.

**Do not** use your textbook or any other reference.

You will not necessarily need to form 10 phylum groups. The number of groups can vary depending on the characteristics that you use to identify the phyla.

Phylum Level

1. Exoskeleton and jointed appendages

2. Endoskeleton

3.

4.

5.

6.

7.

8.

9.

10.

When you complete this portion, have your *instructor check* your groups before you proceed. The first two phyla groups have been selected for you to subdivide into classes. Again *describe* but do not name the group.

Name _____   Section _____

Classes for exoskeletons and jointed legs.
1.
2.
3.
4.
5.
6.

Classes for endoskeletons.
1.
2.
3.
4.
5.
6.

When you have completed this exercise, have it checked by the instructor *before proceeding*.

**Part V: Use of Keys for Phyla and Classes**

The type of key that will be utilized is referred to as a dichotomous or branching key, which means at every level of the key you always have at least two choices. Read the description and decide which one best fits the specimen you are keying. To the right of each description you are referred by number to the next part of the key or the classification name of the phylum or class is given.

Classify each specimen. Check with your instructor to make sure you have classified everything correctly. If any are wrong go through the key again to locate your error. This will be valuable practice for the practical quiz that you will be taking. The *practical quiz* will consist of 20 unknown specimens that you will have to identify to kingdom, phylum, and class *without the use of the key*.

Name _____     Section _____

## A Key of Phyla of the Kingdom Animalia

| | | |
|---|---|---|
| 1a. | Body lacks symmetry; body porous . . . . . . . . . . . . . . . . . . . . . . . . . . . . . . . . . . . . | **Phylum Porifera** |
| 1b. | Body with radial symmetry . . . . . . . . . . . . . . . . . . . . . . . . . . . . . . . . . . . . . . . . . . . . . . | 2 |
| 1c. | Body with bilateral symmetry . . . . . . . . . . . . . . . . . . . . . . . . . . . . . . . . . . . . . . . . . . . . | 3 |
| 2a. | Body soft; disc, tube or saclike with radiating tentacles . . . . . . . . . . . . . . . . . . . . | **Phylum Cnidaria** |
| 2b. | Body hard, spiny or leathery; body parts may be in fives . . . . . . . . . . . . . . . | **Phylum Echinodermata** |
| 3a. | Endoskeletons with backbones or cartilage; appendages when present of two pairs . . . . . . . . . . . . . . . . . . . . . . . **Phylum Chordata (Subphylum Vertebrata)** (go to Class Key) | |
| 3b. | Exoskeletons; appendages, when present, of three or more pairs . . . . . . . . . . . . . . . . . . . . . | 4 |
| 3c. | No skeleton or legs . . . . . . . . . . . . . . . . . . . . . . . . . . . . . . . . . . . . . . . . . . . . . . . . . . . . . . . . . | 4 |
| 4a. | Jointed legs; exoskeletons . . . . . . . . . . . . . . . . . . . . . . . . . | **Phylum Arthropoda** (go to Class Key) |
| 4b. | Without jointed legs; no-skeleton, but shell may be present . . . . . . . . . . . . . . . . . . . . . | 5 |
| 5a. | Body many times longer than wide, but no shell . . . . . . . . . . . . . . . . . . . . . . . . . . . . . . . | 6 |
| 5b. | Body not long and slender, but usually enclosed in a shell; protrusible structure used for locomotion . . . . . . . . . . . . . . . . . . . . . . . . . | **Phylum Mollusca** |
| 6a. | Body flat; with or without segmentation . . . . . . . . . . . . . . . . . . . . . . . . . | **Phylum Platyhelminthes** |
| 6b. | Body round when viewed from end . . . . . . . . . . . . . . . . . . . . . . . . . . . . . . . . . . . . . . . . . | 7 |
| 7a. | Body with segmentation . . . . . . . . . . . . . . . . . . . . . . . . . . . . . . . . . . . . . . . . . . | **Phylum Annelida** |
| 7b. | Body without segmentation . . . . . . . . . . . . . . . . . . . . . . . . . . . . . . . . . . . . . . . | **Phylum Nematoda** |

## A Key of Classes of the Phylum Arthropoda

| | | |
|---|---|---|
| 1a. | Paired antennae present . . . . . . . . . . . . . . . . . . . . . . . . . . . . . . . . . . . . . . . . . . . . . . . . . . . . . | 2 |
| 1b. | Paired antennae absent; 4-5 pair of walking legs . . . . . . . . . . . . . . . . . . . . . . | **Class Arachnida** |
| 2a. | One pair antennae . . . . . . . . . . . . . . . . . . . . . . . . . . . . . . . . . . . . . . . . . . . . . . . . . . . . . . . . . . | 3 |
| 2b. | Two pair antennae; one long, one short . . . . . . . . . . . . . . . . . . . . . . . . . . . . . . . | **Class Crustacea** |
| 3a. | Three pair walking legs . . . . . . . . . . . . . . . . . . . . . . . . . . . . . . . . . . . . . . . . . . . . . . . | **Class Insecta** |
| 3b. | More than three pair walking legs . . . . . . . . . . . . . . . . . . . . . . . . . . . . . . . . . . . . | **Class Myriapoda** |

## A Key of Classes of the Phylum Chordata, Subphylum Vertebrata

| | | |
|---|---|---|
| 1a. | Scales and fins . . . . . . . . . . . . . . . . . . . . . . . . . . . . . . . . . . . . . . . . . . . . . . . . . . . | **Class Osteichthyes** |
| 1b. | No fins, legs usually present . . . . . . . . . . . . . . . . . . . . . . . . . . . . . . . . . . . . . . . . . . . . . . . . | 2 |
| 2a. | Body covering soft and moist; if feet, without claws . . . . . . . . . . . . . . . . . . . . . . | **Class Amphibia** |
| 2b. | Body covering dry and scaly; if feet, with claws . . . . . . . . . . . . . . . . . . . . . . . . . | **Class Reptilia** |
| 2c. | Body covering feathers . . . . . . . . . . . . . . . . . . . . . . . . . . . . . . . . . . . . . . . . . . . . . . . . . | **Class Aves** |
| 2d. | Body covering hair . . . . . . . . . . . . . . . . . . . . . . . . . . . . . . . . . . . . . . . . . . . . . . . . | **Class Mammalia** |

Name _____   Section _____

| Common Name | Phylum | Class (if appropriate) |
|---|---|---|
| 1. Shrimp | | |
| 2. Sand dollar | | |
| 3. Sea urchin | | |
| 4. Planaria | | |
| 5. Roundworm | | |
| 6. Crayfish | | |
| 7. Jellyfish | | |
| 8. Mouse | | |
| 9. Fish | | |
| 10. Beetle | | |
| 11. Cockroach | | |
| 12. Crab | | |
| 13. Sea anemone | | |
| 14. Razor clam | | |
| 15. Hydra | | |
| 16. Bat | | |
| 17. Spider | | |
| 18. Bird | | |
| 19. Grasshopper | | |
| 20. Salamander | | |
| 21. Jellyfish | | |
| 22. Lizard | | |

Name _____     Section _____

23. Sea urchin  _____     _____

24. Oyster  _____     _____

25. Earthworm  _____     _____

26. Chiton  _____     _____

27. Sponge  _____     _____

28. Scorpion  _____     _____

29. Leech  _____     _____

30. Long neck clam  _____     _____

31. Liver fluke  _____     _____

32. Turtle  _____     _____

33. Portugese man-of-war _____     _____

34. Frog  _____     _____

35. Sea star  _____     _____

36. Roundworm  _____     _____

37. Scallop  _____     _____

38. Brittle star  _____     _____

39. Millipede  _____     _____

40. Snake  _____     _____

41. Centipede  _____     _____

42. Snail  _____     _____

43. Horseshoe crab  _____     _____

44. Tapeworm  _____     _____

45. Sandworm  _____     _____

Name _____   Section _____

# LAB 22

# Electrocardiogram and Blood Pressure

**Problem**

How does heart beat, blood pressure and respiration relate in maintaining homeostasis in humans?

**Objectives**

After completing this lab, the student will be able to:

1. Identify the three main parts of a normal EKG (electrocardiogram).
2. Observe effects of muscular contractions, varying respiration, and exercise on EKG.
3. Measure vital capacity using a spirometer.
4. Measure pulse.
5. Determine human blood pressure through the use of a sphygmomanometer and stethoscope.
6. Construct tables and graphs from data obtained from blood pressure and pulse measurements.

**Preliminary Information**

When the heart contracts, blood is pumped to the lungs where carbon dioxide is given off and oxygen is picked up. The carbon dioxide is expelled and new air is drawn into the lungs by the respiratory movements of the diaphragm and muscles which move the ribs.

When the hearts contracts, blood is pumped under pressure to all parts of the body. Blood pressure can be used as an important diagnostic tool and as an indicator of general physical health.

When the heart contracts, electric currents spread into and over the heart. Some of these electric currents spread over the surface of the body and can be recorded from electrodes placed on the body.

The electrocardiograph is an instrument that amplifies and records the voltages produced by the heart as it contracts and relaxes. The electrocardiogram (EKG) is the record that is produced by the electrocardiograph.

**Part I: Electrocardiogram—EKG (ECG)**

The typical EKG records P, QRS, and T waves. The **P wave** represents contraction (excitation) of the atria; **QRS wave** indicates contraction (excitation or depolarization) of the ventricles; the **T wave** indicates relaxation (repolarization) of the ventricles.

By analyzing these wave forms, certain deductions can be made which offer enormous information about the heart.

Name _____  Section _____

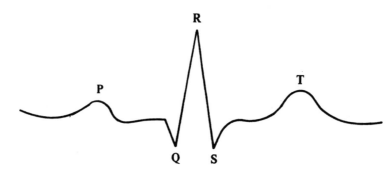

**Figure 22.1** Configuration of the Normal EKG

## Procedure for EKG

1. Have the subject sit in a comfortable position and remove all jewelry.
2. Using alcohol and cotton or an alcohol pad, clean the skin at the sites of electrode attachment (described below), because natural oils of the skin act as insulation.
3. Open one of the packages of electrolyte pads. These pads are damp with a solution allowing good electrical conduction between skin and electrode.
4. Place one pad on each electrode indicated below then **attach the electrode over a bone** near the site indicated on the back of the electrode:
   RA—attach to wrist of right arm
   LA—attach to wrist of left arm
   RL—attach to right leg
   LL—attach to left leg.
5. Electrocardiograph adjustments:
   a. Turn main power switch on.
   b. Set the sensitivity switch on 1.
   c. Set the lead control switch on 2.
   d. Allow the machine to warm up for a minute before proceeding.
   e. To begin taking the EKG, move the center control lever to Run 25.
   NOTE: Abnormal readings indicate an electrode may not be firmly attached.
6. Normal EKG: Have the subject sit quietly to obtain the normal EKG pattern. Run the electrocardiograph for about 20 seconds.
7. Effect of muscle contraction: While keeping other muscles relaxed, have the subject clench one fist.
8. On the electrocardiogram, label:
   a. Your name
   b. P wave
   c. QRS wave
   d. T wave
   e. possible abnormalisites with a circle
   f. fist contraction.
9. Attach the EKG strip to the data sheet.

Name _____  Section _____

## Part II: Pulmonary Function

Air is drawn into the lungs by a combination of the action of muscles between the ribs and the diaphragm muscles across the abdomen. The physical dimensions of your lungs vary according to heredity, sex, age, smoking, exercise and other factors.

The air in your lungs can be defined in the following way:

- Tidal Volume: amount of air inspired or expired with each normal breath.
- Expiratory Reserve Volume: air that can still be expired by forceful expiration after the end of normal tidal expiration.
- Vital Capacity: the maximum volume of air that can be exhaled after taking a deep breath.

1. Put a clean disposable cardboard mouth piece on the spirometer.
2. Turn the face of the spirometer until the red zero mark lines up with the needle.
3. Air exhaled through the spirometer is recorded in cubic centimeters (cc) by the movement of the needle.
4. To measure vital capacity
   a. Inhale as much as you can,
   b. hold you nose closed,
   c. exhale as much as you can through the spirometer.
5. Compare your resuls with the "Percentage of Vital Capacity Chart."
6. Record the data in the table on the data sheet.

## Part III: Pulse Rate Measurement

1. Working in groups of two, determine each others pulse rate by placing two fingers (not a thumb) over the main artery in the wrist, located just below the upper bone in the forearm.
   Once you have located the artery and felt the pulse, count the pulse rate for 60 seconds. Record your pulse rate on your data sheet.
   **Optional:**
2. Working in either small groups or as a class, measure the pulse rate of a smoker both before and after smoking a cigarette. Step out of the building to smoke. Record your findings on the data sheet.

## Part IV: Human Blood Pressure Determinations by Sphygmomanometers

Contraction of the ventricles exerts pressure on the blood vessels and is called systole. **Systolic pressure** is the pressure of the blood during ventricular contraction. This pressure **normally ranges between 110-140.** Following contraction the heart relaxes and blood rushes in from the pulmonary veins and venae cavae, filling the atria. This is diastole and this lower pressure is the **diastolic pressure.** This **ranges normally from 60-90.**

Name _____  Section _____

A. Apparatus

A sphygmomanometer consists of (1) a compression bag surrounded by an unyielding cuff for application of an extra-arterial pressure, (2) an inflating bulb, pump, or other device by which pressure is created in the system, and (3) a manometer by which the applied pressure is read, (4) a variable, controllable exhaust by which the system can be deflated either gradually or rapidly.

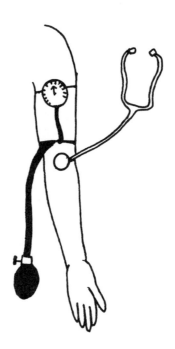

**Figure 22.2** Position of Sphygmomanometer and Stethoscope

B. Technique

1. Read and understand each point before beginning. The sphygmomanometer should be inflated only about 45 seconds or less.

2. The patient. The subject should be comfortably seated. The arm should be bared, slightly flexed, and perfectly relaxed. In the sitting position the forearm should be supported at heart level. Keep the arm off the table to avoid hearing artifact noises. The deflated bag and cuff should be applied evenly and snugly around the arm with the lower edge about 3 cm above the antecubital space (inside of elbow) with the rubber tubing attached to the pressure gauge in line with the antecubital space. Anyone who has high blood pressure should not participate in the experiment.

3. Determination of systolic pressure. Clean stethoscope ear pieces with alcohol and cotton or alcohol pad to prevent possible ear infection. Put the stethoscope on and allow a minute to familiarize yourself with the sounds heard through the stethoscope by listening to your own heart.

    In taking a blood pressure, the stethoscope receiver should be applied snugly over the artery in the antecubital space, free from contact with the cuff. See Figure 22.2. The pressure in the sphygmomanometer should then be raised rapidly to about 100 by tightening the screw and pumping up the cuff. Allow a minute to familiarize yourself with the sound of the heartbeat through the artery in the elbow. Now rapidly pump up to 160. Decrease *slowly* by slightly loosening the screw until a sound is heard with each heartbeat. Note the reading on the pressure gauge and record in space below as systolic pressure.

Name _____ Section _____

4. Determination of diastolic pressure. With continued deflation of the system below systolic pressure at a rate of 2 to 3 per heartbeat, the sounds undergo changes in intensity and quality. As the cuff pressure approaches the diastolic, the sounds often become dull and muffled quite suddenly and finally cease. The point of complete cessation is the best index of diastolic pressure. Record the subject's diastolic pressure.
5. Recording blood pressure. Blood pressure is recorded with the systolic over the diastolic:

    118/74 is 118 over 74. Record your blood pressure on data sheet.

6. Rotate members of your lab group so that each of you will have an opportunity to use the apparatus.
7. Variability in blood pressure.
    Exercise test (three students may work together on this experiment)
    (1) Determine blood pressure from the one arm and pulse from the other arm with the subject sitting. Record on data sheet.
    (2) Have the subject exercise vigorously for 1-2 minutes.
    (3) Take the blood pressure and pulse immediately after exercise and then every 2 minutes until the readings return to normal or stabilize.
    (4) On your data sheet, record your results and plot a graph.

## Data Sheet

## Part I: Electrocardiogram

Attach your EKG strip here.

What happened to the EKG when a fist was clenched? Explain.

Name _____  Section _____

## Part II: Pulmonary Function

**TABLE 22.1**
Spirometer Data

| Subject | Sex | Height | Vital Capacity | Percent Vital Capacity[1] |
|---------|-----|--------|----------------|---------------------------|
| 1 | | | | |
| 2 | | | | |
| 3 | | | | |
| 4 | | | | |

[1] Use Table 22.2

Name ______________________ Section ______________________

# TABLE 22.2
Percentage of Vital Capacity Chart

**PERCENTAGE OF VITAL CAPACITY MEN (CALCULATED FROM STANDING HEIGHT) IN CUBIC CENTIMETERS**

| Standing height in cms. | ins. | 700 | 800 | 900 | 1000 | 1100 | 1200 | 1300 | 1400 | 1500 | 1600 | 1700 | 1800 | 1900 | 2000 | 2100 | 2200 | 2300 | 2400 | 2500 | 2600 | 2700 | 2800 | 2900 | 3000 | 3100 | 3200 |
|---|---|---|---|---|---|---|---|---|---|---|---|---|---|---|---|---|---|---|---|---|---|---|---|---|---|---|---|
| 144.8 | 57 | 19 | 22 | 25 | 28 | 30 | 33 | 36 | 39 | 41 | 44 | 47 | 50 | 52 | 55 | 58 | 61 | 64 | 66 | 69 | 72 | 75 | 77 | 80 | 83 | 86 | 88 |
| 147.3 | 58 | 19 | 22 | 24 | 27 | 30 | 33 | 35 | 38 | 41 | 43 | 46 | 49 | 52 | 54 | 57 | 60 | 63 | 65 | 68 | 71 | 73 | 76 | 79 | 82 | 84 | 87 |
| 149.8 | 59 | | 21 | 24 | 27 | 30 | 32 | 35 | 37 | 40 | 43 | 45 | 48 | 51 | 53 | 56 | 59 | 61 | 64 | 67 | 69 | 72 | 75 | 77 | 80 | 83 | 85 |
| 152.4 | 60 | | 21 | 24 | 26 | 29 | 32 | 34 | 37 | 39 | 42 | 45 | 47 | 50 | 53 | 55 | 58 | 60 | 63 | 66 | 68 | 71 | 74 | 76 | 79 | 81 | 84 |
| 154.9 | 61 | | 21 | 23 | 26 | 28 | 31 | 34 | 36 | 39 | 41 | 44 | 47 | 49 | 52 | 54 | 57 | 59 | 62 | 65 | 67 | 70 | 72 | 75 | 78 | 80 | 83 |
| 157.4 | 62 | | 20 | 23 | 25 | 28 | 30 | 33 | 36 | 38 | 41 | 43 | 46 | 48 | 51 | 53 | 56 | 58 | 61 | 64 | 66 | 69 | 71 | 74 | 76 | 79 | 81 |
| 160.0 | 63 | | 20 | 23 | 25 | 28 | 30 | 33 | 35 | 38 | 40 | 43 | 45 | 48 | 50 | 53 | 55 | 58 | 60 | 63 | 65 | 68 | 70 | 73 | 75 | 78 | 80 |
| 162.5 | 64 | | 20 | 22 | 25 | 27 | 30 | 32 | 35 | 37 | 39 | 42 | 44 | 47 | 49 | 52 | 54 | 57 | 59 | 62 | 64 | 67 | 69 | 71 | 74 | 76 | 79 |
| 165.1 | 65 | | | 22 | 24 | 27 | 29 | 32 | 34 | 36 | 39 | 41 | 44 | 46 | 48 | 51 | 53 | 56 | 58 | 61 | 63 | 66 | 68 | 70 | 73 | 75 | 78 |
| 167.6 | 66 | | | 21 | 24 | 26 | 29 | 31 | 33 | 36 | 38 | 41 | 43 | 45 | 48 | 50 | 52 | 55 | 57 | 60 | 62 | 64 | 67 | 69 | 71 | 74 | 76 |
| 170.2 | 67 | | | 21 | 24 | 26 | 28 | 31 | 33 | 35 | 38 | 40 | 42 | 45 | 47 | 49 | 52 | 54 | 56 | 59 | 61 | 63 | 66 | 68 | 71 | 73 | 75 |
| 172.7 | 68 | | | 21 | 23 | 26 | 28 | 30 | 32 | 35 | 37 | 39 | 42 | 44 | 46 | 49 | 51 | 53 | 56 | 58 | 60 | 63 | 65 | 67 | 70 | 72 | 74 |
| 175.5 | 69 | | | 21 | 23 | 25 | 27 | 30 | 32 | 34 | 37 | 39 | 41 | 43 | 46 | 48 | 50 | 52 | 55 | 57 | 59 | 62 | 64 | 66 | 68 | 71 | 73 |
| 177.8 | 70 | | | 20 | 22 | 25 | 27 | 29 | 31 | 34 | 36 | 38 | 40 | 43 | 45 | 47 | 49 | 52 | 54 | 56 | 58 | 61 | 63 | 65 | 67 | 70 | 72 |
| 180.4 | 71 | | | 20 | 22 | 24 | 27 | 29 | 31 | 33 | 35 | 38 | 40 | 42 | 44 | 47 | 49 | 51 | 53 | 55 | 58 | 60 | 62 | 64 | 67 | 69 | 71 |
| 182.9 | 72 | | | 20 | 22 | 24 | 26 | 28 | 31 | 33 | 35 | 37 | 39 | 42 | 44 | 46 | 48 | 50 | 53 | 55 | 57 | 59 | 61 | 63 | 66 | 68 | 70 |
| 185.4 | 73 | | | 19 | 22 | 24 | 26 | 28 | 30 | 32 | 35 | 37 | 39 | 41 | 43 | 45 | 47 | 50 | 52 | 54 | 56 | 58 | 60 | 63 | 65 | 67 | 69 |
| 188.0 | 74 | | | 19 | 21 | 23 | 26 | 28 | 30 | 32 | 34 | 36 | 38 | 40 | 43 | 45 | 47 | 49 | 51 | 53 | 55 | 57 | 60 | 62 | 64 | 66 | 68 |
| 190.5 | 75 | | | | 21 | 23 | 25 | 27 | 29 | 32 | 34 | 36 | 38 | 40 | 42 | 44 | 46 | 48 | 50 | 53 | 55 | 57 | 59 | 61 | 63 | 65 | 67 |
| 193.0 | 76 | | | | 21 | 23 | 25 | 27 | 29 | 31 | 33 | 35 | 37 | 39 | 42 | 44 | 46 | 48 | 50 | 52 | 54 | 56 | 58 | 60 | 62 | 64 | 66 |
| | | 42.7 | 48.8 | 54.9 | 61.0 | 62.1 | 73.2 | 79.3 | 85.4 | 91.5 | 97.6 | 103.7 | 109.8 | 115.9 | 122.0 | 128.1 | 134.2 | 140.3 | 146.4 | 152.5 | 158.6 | 164.7 | 170.8 | 176.9 | 183.0 | 189.1 | 195.2 |

| Standing height in cms. | ins. | 3300 | 3400 | 3500 | 3600 | 3700 | 3800 | 3900 | 4000 | 4100 | 4200 | 4300 | 4400 | 4500 | 4600 | 4700 | 4800 | 4900 | 5000 | 5100 | 5200 | 5300 | 5400 | 5500 | 5600 | 5700 | 5800 |
|---|---|---|---|---|---|---|---|---|---|---|---|---|---|---|---|---|---|---|---|---|---|---|---|---|---|---|---|
| 144.8 | 57 | 91 | 94 | 97 | 99 | 102 | 105 | 108 | 110 | 113 | 116 | 119 | 121 | | | | | | | | | | | | | | |
| 147.3 | 58 | 90 | 92 | 95 | 98 | 100 | 103 | 106 | 109 | 111 | 114 | 117 | 120 | 122 | | | | | | | | | | | | | |
| 149.8 | 59 | 88 | 91 | 93 | 96 | 99 | 101 | 104 | 107 | 109 | 112 | 115 | 117 | 120 | 123 | | | | | | | | | | | | |
| 152.4 | 60 | 87 | 89 | 92 | 95 | 97 | 100 | 102 | 105 | 108 | 110 | 113 | 115 | 118 | 121 | 123 | | | | | | | | | | | |
| 154.9 | 61 | 85 | 88 | 90 | 93 | 96 | 98 | 101 | 103 | 106 | 108 | 111 | 114 | 116 | 119 | 121 | 124 | | | | | | | | | | |
| 157.4 | 62 | 84 | 86 | 89 | 91 | 94 | 97 | 99 | 102 | 104 | 107 | 109 | 112 | 114 | 117 | 119 | 122 | | | | | | | | | | |
| 160.0 | 63 | 83 | 85 | 88 | 90 | 93 | 95 | 98 | 100 | 103 | 105 | 108 | 110 | 113 | 115 | 118 | 120 | 123 | | | | | | | | | |
| 162.5 | 64 | 81 | 84 | 86 | 89 | 91 | 94 | 96 | 99 | 101 | 103 | 106 | 108 | 111 | 113 | 116 | 118 | 120 | 123 | | | | | | | | |
| 165.1 | 65 | 80 | 82 | 85 | 87 | 90 | 92 | 95 | 97 | 99 | 102 | 104 | 107 | 109 | 111 | 114 | 116 | 119 | 121 | 123 | | | | | | | |
| 167.6 | 66 | 79 | 81 | 83 | 86 | 88 | 91 | 93 | 95 | 98 | 100 | 102 | 105 | 107 | 110 | 112 | 114 | 117 | 119 | 121 | 124 | | | | | | |
| 170.2 | 67 | 78 | 80 | 82 | 85 | 87 | 89 | 92 | 94 | 97 | 99 | 101 | 104 | 106 | 108 | 111 | 113 | 115 | 118 | 120 | 122 | | | | | | |
| 172.7 | 68 | 76 | 79 | 81 | 83 | 86 | 88 | 90 | 93 | 95 | 97 | 100 | 102 | 104 | 107 | 109 | 111 | 114 | 116 | 118 | 120 | 123 | | | | | |
| 175.5 | 69 | 75 | 78 | 80 | 82 | 84 | 87 | 89 | 91 | 94 | 96 | 98 | 100 | 103 | 105 | 107 | 109 | 112 | 114 | 116 | 119 | 121 | 123 | | | | |
| 177.8 | 70 | 74 | 76 | 79 | 81 | 83 | 86 | 88 | 90 | 92 | 94 | 97 | 99 | 101 | 103 | 106 | 108 | 110 | 112 | 115 | 117 | 119 | 121 | 124 | | | |
| 180.4 | 71 | 73 | 75 | 78 | 80 | 82 | 84 | 86 | 89 | 91 | 93 | 95 | 98 | 100 | 102 | 104 | 106 | 109 | 111 | 113 | 115 | 117 | 120 | 122 | 124 | | |
| 182.9 | 72 | 72 | 74 | 77 | 79 | 81 | 83 | 85 | 88 | 90 | 92 | 94 | 96 | 98 | 101 | 103 | 105 | 107 | 109 | 111 | 114 | 116 | 118 | 120 | 122 | 125 | |
| 185.4 | 73 | 71 | 73 | 76 | 78 | 80 | 82 | 84 | 86 | 89 | 91 | 93 | 95 | 97 | 99 | 101 | 103 | 106 | 108 | 110 | 112 | 114 | 117 | 119 | 121 | 123 | |
| 188.0 | 74 | 70 | 72 | 75 | 77 | 79 | 81 | 83 | 85 | 87 | 89 | 92 | 94 | 96 | 98 | 100 | 102 | 104 | 106 | 108 | 111 | 113 | 116 | 117 | 119 | 121 | 123 |
| 190.5 | 75 | 69 | 71 | 73 | 76 | 78 | 80 | 82 | 84 | 86 | 88 | 90 | 92 | 95 | 97 | 99 | 101 | 103 | 105 | 107 | 109 | 111 | 113 | 116 | 118 | 120 | 122 |
| 193.0 | 76 | 68 | 70 | 72 | 75 | 77 | 79 | 81 | 83 | 85 | 87 | 89 | 91 | 93 | 95 | 97 | 99 | 101 | 104 | 106 | 108 | 110 | 112 | 114 | 116 | 118 | 120 |
| | | 201.3 | 207.4 | 213.5 | 219.6 | 225.7 | 231.8 | 237.9 | 244.0 | 250.1 | 256.2 | 262.3 | 268.4 | 274.5 | 280.6 | 286.7 | 292.8 | 298.9 | 305.0 | 311.1 | 317.2 | 323.3 | 329.4 | 335.5 | 341.6 | 347.7 | 353.8 |

**PERCENTAGE OF VITAL CAPACITY IN CUBIC INCHES**

**PERCENTAGE OF VITAL CAPACITY WOMEN (CALCULATED FROM STANDING HEIGHT) IN CUBIC CENTIMETERS**

| Standing Height in cm. | Standing Height in inches | 600 | 700 | 800 | 900 | 1000 | 1100 | 1200 | 1300 | 1400 | 1500 | 1600 | 1700 | 1800 | 1900 | 2000 | 2100 | 2200 | 2300 | 2400 | 2500 |
|---|---|---|---|---|---|---|---|---|---|---|---|---|---|---|---|---|---|---|---|---|---|
| 139.7 | 55 | 21 | 23 | 29 | 32 | 36 | 39 | 43 | 47 | 50 | 54 | 57 | 61 | 64 | 68 | 71 | 75 | 79 | 82 | 86 | 89 |
| 142.2 | 56 | 21 | 23 | 28 | 32 | 33 | 39 | 42 | 46 | 49 | 53 | 56 | 60 | 63 | 67 | 70 | 74 | 77 | 81 | 84 | 88 |
| 144.8 | 57 | 21 | 24 | 28 | 31 | 35 | 38 | 41 | 45 | 48 | 52 | 55 | 59 | 62 | 66 | 69 | 72 | 76 | 79 | 83 | 86 |
| 147.3 | 58 | 20 | 24 | 27 | 31 | 34 | 38 | 41 | 44 | 47 | 51 | 54 | 58 | 61 | 64 | 68 | 71 | 75 | 78 | 81 | 85 |
| 149.8 | 59 | 20 | 23 | 27 | 30 | 33 | 37 | 40 | 43 | 47 | 50 | 53 | 57 | 60 | 63 | 67 | 70 | 73 | 77 | 80 | 83 |
| 152.4 | 60 | 20 | 23 | 26 | 30 | 33 | 36 | 39 | 43 | 46 | 49 | 52 | 56 | 59 | 62 | 66 | 69 | 72 | 75 | 79 | 82 |
| 154.9 | 61 | | 23 | 26 | 29 | 32 | 36 | 39 | 42 | 45 | 48 | 52 | 55 | 58 | 61 | 65 | 68 | 71 | 74 | 77 | 81 |
| 157.4 | 62 | | 22 | 25 | 29 | 32 | 35 | 36 | 41 | 44 | 48 | 51 | 54 | 57 | 60 | 63 | 67 | 70 | 73 | 76 | 79 |
| 160.0 | 63 | | 22 | 25 | 28 | 31 | 35 | 38 | 41 | 44 | 47 | 50 | 53 | 56 | 59 | 62 | 66 | 69 | 72 | 75 | 78 |
| 162.5 | 64 | | 22 | 25 | 28 | 31 | 34 | 37 | 40 | 43 | 46 | 49 | 52 | 55 | 58 | 62 | 65 | 68 | 71 | 74 | 77 |
| 165.1 | 65 | | 21 | 24 | 27 | 30 | 33 | 36 | 39 | 42 | 46 | 49 | 52 | 55 | 58 | 61 | 64 | 67 | 70 | 73 | 76 |
| 167.6 | 66 | | 21 | 24 | 27 | 30 | 33 | 36 | 39 | 42 | 45 | 48 | 51 | 54 | 57 | 60 | 63 | 66 | 69 | 72 | 75 |
| 170.2 | 67 | | 21 | 24 | 27 | 29 | 32 | 35 | 38 | 41 | 44 | 47 | 50 | 53 | 56 | 59 | 62 | 65 | 68 | 71 | 74 |
| 172.7 | 68 | | 20 | 23 | 26 | 29 | 32 | 35 | 38 | 41 | 44 | 46 | 49 | 52 | 55 | 58 | 61 | 64 | 67 | 69 | 72 |
| 175.3 | 69 | | 20 | 23 | 26 | 29 | 31 | 34 | 37 | 40 | 43 | 46 | 49 | 51 | 54 | 57 | 60 | 63 | 66 | 68 | 71 |
| 177.8 | 70 | | 20 | 22 | 25 | 28 | 31 | 34 | 37 | 39 | 42 | 45 | 48 | 51 | 53 | 56 | 59 | 62 | 65 | 67 | 70 |
| 180.4 | 71 | | | 22 | 25 | 28 | 31 | 33 | 36 | 39 | 42 | 44 | 47 | 50 | 53 | 55 | 58 | 61 | 64 | 66 | 69 |
| 182.9 | 72 | | | 22 | 25 | 27 | 30 | 33 | 36 | 38 | 41 | 44 | 47 | 49 | 52 | 55 | 57 | 60 | 63 | 66 | 68 |
| 185.4 | 73 | | | 22 | 24 | 27 | 30 | 32 | 35 | 38 | 40 | 43 | 46 | 49 | 51 | 54 | 57 | 59 | 62 | 65 | 67 |
| 188.0 | 74 | | | 21 | 24 | 27 | 29 | 32 | 35 | 37 | 40 | 43 | 45 | 48 | 51 | 53 | 56 | 59 | 61 | 64 | 67 |
| | | 36.6 | 42.7 | 48.8 | 54.9 | 61.0 | 62.1 | 73.2 | 79.3 | 85.4 | 91.5 | 97.6 | 103.7 | 109.8 | 115.9 | 122.0 | 128.1 | 134.2 | 140.3 | 146.4 | 152.5 |

| Standing Height in cm. | Standing Height in inches | 2600 | 2700 | 2800 | 2900 | 3000 | 3100 | 3200 | 3300 | 3400 | 3500 | 3600 | 3700 | 3800 | 3900 | 4000 | 4100 | 4200 | 4300 | 4400 | 4500 |
|---|---|---|---|---|---|---|---|---|---|---|---|---|---|---|---|---|---|---|---|---|---|
| 139.7 | 55 | 93 | 96 | 100 | 103 | 107 | 111 | 114 | 118 | 121 | | | | | | | | | | | |
| 142.2 | 56 | 91 | 95 | 98 | 102 | 105 | 109 | 112 | 116 | 119 | 123 | | | | | | | | | | |
| 144.8 | 57 | 90 | 93 | 97 | 100 | 103 | 107 | 110 | 114 | 117 | 121 | | | | | | | | | | |
| 147.3 | 58 | 88 | 91 | 95 | 98 | 102 | 105 | 108 | 112 | 115 | 118 | 122 | | | | | | | | | |
| 149.8 | 59 | 87 | 90 | 93 | 97 | 100 | 103 | 107 | 110 | 113 | 117 | 120 | 120 | | | | | | | | |
| 152.4 | 60 | 85 | 88 | 92 | 95 | 98 | 102 | 105 | 108 | 112 | 115 | 118 | 121 | | | | | | | | |
| 154.9 | 61 | 84 | 87 | 90 | 94 | 97 | 100 | 103 | 106 | 110 | 113 | 116 | 119 | 122 | | | | | | | |
| 157.4 | 62 | 83 | 86 | 89 | 92 | 95 | 98 | 102 | 105 | 108 | 111 | 114 | 118 | 121 | | | | | | | |
| 160.0 | 63 | 81 | 84 | 88 | 91 | 94 | 97 | 100 | 103 | 106 | 109 | 112 | 116 | 119 | 122 | | | | | | |
| 162.5 | 64 | 80 | 83 | 86 | 89 | 92 | 95 | 98 | 101 | 105 | 108 | 111 | 114 | 117 | 120 | | | | | | |
| 165.1 | 65 | 79 | 82 | 85 | 88 | 91 | 94 | 97 | 100 | 103 | 106 | 109 | 112 | 115 | 118 | 121 | | | | | |
| 167.6 | 66 | 78 | 81 | 84 | 87 | 90 | 93 | 96 | 98 | 101 | 104 | 107 | 110 | 113 | 116 | 119 | 122 | | | | |
| 170.2 | 67 | 77 | 79 | 82 | 85 | 88 | 91 | 94 | 97 | 100 | 103 | 106 | 109 | 112 | 115 | 118 | 120 | 123 | | | |
| 172.7 | 68 | 75 | 78 | 81 | 84 | 87 | 90 | 93 | 96 | 98 | 101 | 104 | 107 | 110 | 113 | 116 | 119 | 121 | | | |
| 175.3 | 69 | 74 | 77 | 80 | 83 | 85 | 88 | 91 | 94 | 97 | 100 | 102 | 105 | 108 | 111 | 114 | 117 | 120 | 123 | | |
| 177.8 | 70 | 73 | 76 | 79 | 82 | 84 | 87 | 90 | 93 | 96 | 98 | 101 | 104 | 107 | 110 | 112 | 115 | 118 | 121 | | |
| 180.4 | 71 | 72 | 75 | 78 | 80 | 83 | 86 | 89 | 91 | 94 | 97 | 100 | 102 | 105 | 108 | 111 | 114 | 116 | 119 | 122 | |
| 182.9 | 72 | 71 | 74 | 77 | 79 | 82 | 85 | 88 | 90 | 93 | 96 | 98 | 101 | 104 | 107 | 109 | 112 | 115 | 118 | 120 | |
| 185.4 | 73 | 70 | 73 | 76 | 78 | 81 | 84 | 86 | 89 | 92 | 94 | 97 | 100 | 102 | 105 | 108 | 111 | 113 | 116 | 119 | 121 |
| 188.0 | 74 | 69 | 72 | 75 | 77 | 80 | 83 | 85 | 88 | 90 | 93 | 96 | 99 | 101 | 104 | 106 | 109 | 112 | 114 | 117 | 120 |
| | | 158.6 | 164.7 | 170.8 | 176.9 | 183.0 | 189.1 | 195.2 | 201.3 | 207.4 | 213.5 | 219.6 | 225.7 | 231.8 | 237.9 | 244.0 | 250.1 | 256.2 | 262.3 | 268.4 | 274.5 |

**PERCENTAGE OF VITAL CAPACITY IN CUBIC INCHES**

Name _____   Section _____

## Part III: Pulse Rate

Your pulse rate _____

Smoker's pulse rate:   Normal _____

After smoking _____

## Part IV: Human Blood Pressure Determination

Your normal blood pressure _____

**TABLE 22.3**
Blood Pressure and Pulse Following Exercise

|  |  | Normal | Immediately | Time After Exercise ||||| 
|---|---|---|---|---|---|---|---|---|
|  |  |  |  | 2 Min | 4 Min | 6 Min. | 8 Min. | 10 Min. |
| Blood Pressure | Systolic |  |  |  |  |  |  |  |
|  | Diastolic |  |  |  |  |  |  |  |
| Pulse |  |  |  |  |  |  |  |  |

Name _____   Section _____

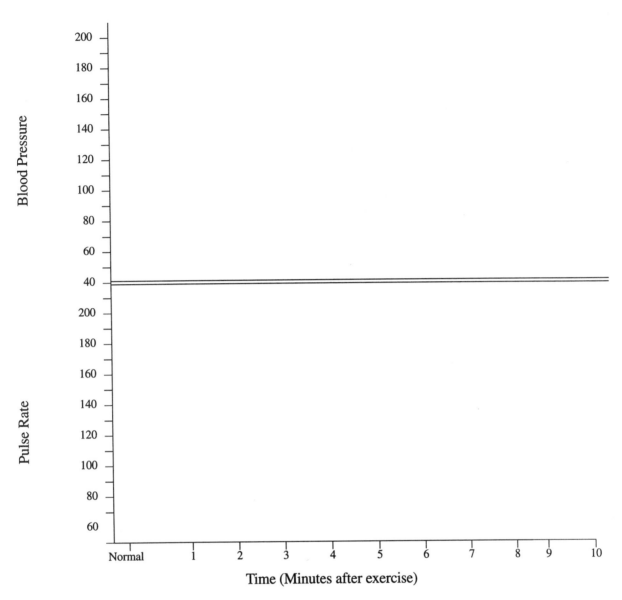

**Figure 22.3** Blood Pressure and Pulse Following Exercise

Graph one line for the systolic pressure, and graph a second line for the disastolic pressure. The third line is a graph of the pulse rate.

Name _____   Section _____

# LAB 23

# Animal Structure: Dissection of the Fetal Pig

*Note:* Terms in bold print should be identified on both the fetal pig and human manikin.

## Problem

What is the appearance and physical relationship of the respiratory, digestive, excretory, reproductive and circulatory systems of the fetal pig and human?

## Objectives

After completing this lab exercise the student will be able to:

1. Take a written or oral laboratory practical quiz identifying the major structures of the respiratory, digestive, excretory, reproductive and circulatory systems of the fetal pig and human. These structures are printed in bold.
2. List the functions of the specified structures.

## Preliminary Information

This laboratory work will deal with the anatomy (structure) and physiology (function) of some parts of the five major organ systems which are most directly concerned with the maintenance of life. The organs and their relationships in the fetal pig are the same as those found in most mammals, including humans.

The pig normally has a 17 week gestation period. The fetal pig you will be studying was removed from its sow by a slaughter house sometime during the last few weeks of gestation, processed by a biological supply company, then purchased by the school.

Whenever you are working with the anatomy of any organism, certain terms are commonly used. Many of these are listed here and you should know and be able to use all of them.

**anterior**—the head region
**posterior**—the tail region or end region
**dorsal**—the back side; up
**ventral**—the belly side; down
**caudal**—the tail region
**lateral**—the right or left side
**medial**—the middle or center
**proximal**—closer to the middle
**distal**—farther from the middle
**pectoral**—the shoulder region
**pelvic**—the hip region
**right and left**—the pig's right and left

Name _____    Section _____

## Part I: External Observation

1. Working in group of two, obtain a dissecting pan, string (or rubber bands), six large dissecting pins, and a plastic bag from the supply area.
2. Obtain a fetal pig from the supply area.
3. After you have finished dissecting, remove all pins, place the pig and string in the plastic bag, tie, label with both partner's names written in pencil on masking tape, and place upright in the receptacle that is provided for your lab section.
4. During the external examination of the pig you are to observe the position of certain anatomical features. Observe these body regions: head, neck, trunk with two pairs of appendages, and tail. As in all mammals, the pig's trunk is divided into a **thorax**—the area from the anterior end of the pectoral region to the posterior end of the rib cage, and **abdomen**—the area from the posterior end of the rib cage to the posterior end of the pelvic region. The thoracic region contains the heart and lungs. The abdominal region contains the digestive, excretory, and reproductive organs.
5. On the ventral surface, there is a large **umbilical cord**. Three blood vessels pass through this cord. Two smaller thick-walled umbilical arteries and one large thin-walled umbilical vein carry oxygen, nutrients and waste products between the fetal pig and its mother's placenta (the area of nutrient and waste exchange between fetal pig and mother). The umbilical arteries carry blood with waste products away from the fetal pig to the placenta. The umbilical vein carries oxygen and nutrients from the placenta to the fetal pig.

## Part II: Organs of the Abdominal Cavity

*Note:* Dissection instructions should be closely followed and no organs should be removed unless directed.

Place the pig ventral side up in the dissecting pan as shown in figure 23.1. To keep the pig in this position, tie the string to one front leg, run the string under the pan and tie it to the other front leg. Do the same for the back legs.

Use a scissors only to open the abdominal and thoracic cavities. While pulling the umbilical cord away from the body, make a left-to-right cut through the body wall one centimeter anterior to the umbilical cord. When the abdominal cavity has been reached, insert one blade of the scissors into the cavity and continue to open the abdominal and thoracic cavities following the pattern in figure 23.1. Cut 1-2 will involve splitting the sternum.

Pick up the end of the umbilical cord and raise it slightly. Notice that it is attached by a vessel to one of the visceral organs. This is the **umbilical vein**. *Tie a string around this vein so it can be located later, and clip the vein between the string and umbilical cord.* Pull the umbilical cord posteriorly and let it rest between the hind legs. The abdominal cavity is now exposed, as in figure 23.2. The structures within the coelom or body cavity should now be gently washed with a stream of tap water to remove any clotted blood.

Cut off the flaps of the body wall on both sides of the pig. Cut through the ribs along the right and left sides of the pig to remove the rib cage. This will allow access to the thoracic cavity. Discard all flaps.

With the aid of figures 23.3 and 23.4 identify and study:

1. The **coelom** or body cavity which is divided into thoracic and abdominal cavities by the thin **diaphragm**.
2. The **peritoneum** lining the abdominal cavity and covering its **viscera** or internal organs. **Mesentery** is the double layer of peritoneum surrounding certain abdominal organs and attaching to the dorsal abdominal wall. Mesenteries are most easily located surrounding and holding the small intestine and its blood vessels.

Name _____     Section _____

3. The **liver**, a reddish-brown four-lobed structure posterior to the diaphragm.
4. The **gallbladder**, a membranous bag or sac the size of a pea partially imbedded in the dorsal side of the right medial lobe of the liver near its posterior border.
5. The **cystic duct** which leads out of the gallbladder, joins the **hepatic ducts** from the liver to form a **common bile duct** that enters the duodenum. The **duodenum** is the first portion of the small intestine.
6. The **stomach** with the esophagus entering the **cardiac** portion and ending with the constructed **pyloric** portion.
7. The **spleen**, a flat elongated reddish-tan structure that lies just to the left of the stomach.
8. The light-colored, granular **pancreas** lying slightly dorsal and posterior to the stomach. Lift the stomach and small intestines to examine the pancreas. Use paper towels to blot up any excess fluids.
9. The much coiled **small intestine** attached to the stomach at the stomach's pyloric sphincter, a muscular valve. Raise the small intestine near the middle and examine the mesentery and blood vessels. The small intestine is usually subdivided into three sections. The **duodenum** is the first 20 cm in an adult, jejunum is the middle section and the last ⅓ is the ilium which connects to the colon.
10. The **large intestine** which follows the small intestine. There are several distinct parts of the large intestine:
    a. **The colon**, the anterior coiled part.
    b. **The rectum**, the somewhat straight posterior portion which opens to the outside through the **anus**.
    c. **The cecum**, the blind saclike beginning of the large intestine, at the juncture of the small and large intestines. In humans, the **appendix** is attached to the cecum.
11. The long **urinary bladder** which extends ventro-posteriorly from the umbilical area. The right and left umbilical arteries lie along the sides of the bladder. In the initial dissection the umbilical cord and urinary bladder were pulled back between the pig's hind legs.
12. The **kidneys**, large bean-shaped organs lying against the dorsal wall in the middle of the abdominal cavity. The kidneys are easily seen by lifting the intestines.
13. The **ureter**, a tube carrying urine from the kidney to the urinary bladder.

## Part III: Organs of the Thoracic Cavity

In the thoracic cavity observe:

1. The **pleural membrane**, the thin shiny membrane lining the cavity and covering the organs in it.
2. The **pericardium**, the sac which surrounds the heart.
3. A portion of the whitish **thymus gland** is attached to the pericardium. The rest of the thymus gland will be located later in the neck region.
4. The **lungs** which lie on each side of the heart.
5. The **esophagus** which can be located anterior to the diaphragm. By tearing much of the pleural membrane between the lobes of the lungs and pulling the lobes to one side, three tubes can be seen: **posterior vena cava**, the most ventral tube; **esophagus**, the middle tube; **dorsal aorta**, the most dorsal tube. Remove and discard the left lung.

Name _____  Section _____

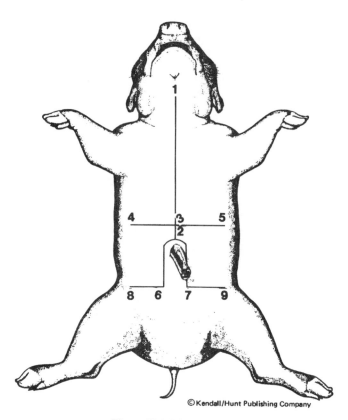

**Figure 23.1** Dissection Cuts

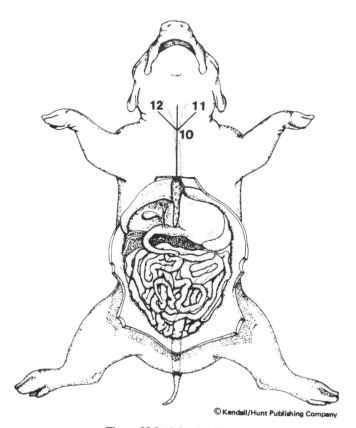

**Figure 23.2** Abdominal Cavity

Name _____ Section _____

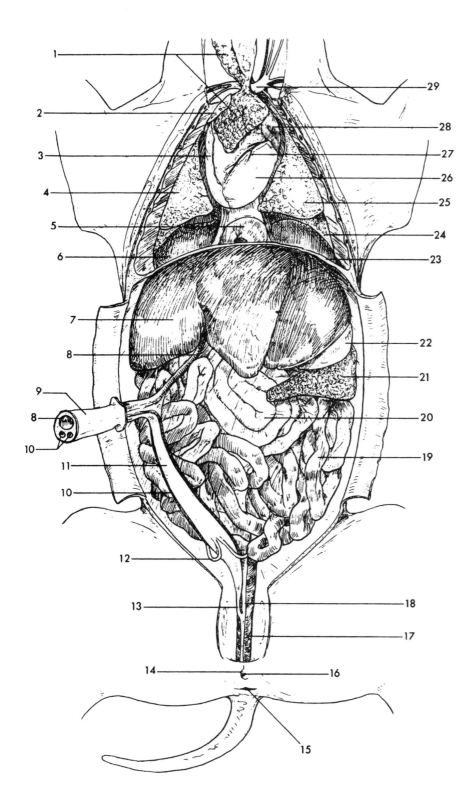

1. Thymus
2. Lung, right apical lobe
3. Pericardium, partly removed
4. Lung, right cardiac lobe
5. Lung, right intermediate lobe
6. Lung, right diaphragmatic lobe
7. Liver
8. Umbilical vein
9. Umbilical cord
10. Umbilical artery
11. Urinary bladder
12. Ureter
13. Urethra
14. Genital papilla
15. Anus
16. External urogenital orifice
17. Rectum
18. Vagina
19. Small intestine
20. Large intestine
21. Spleen
22. Stomach
23. Diaphragm
24. Lung, left diaphragmatic lobe
25. Lung, left cardiac lobe
26. Left ventricle
27. Coronary artery and vein
28. Left atrium
29. Aortic arch

Illustration by Carolina Biological Supply Company, © 1972

**Figure 23.3** Fetal Pig, Internal Organs

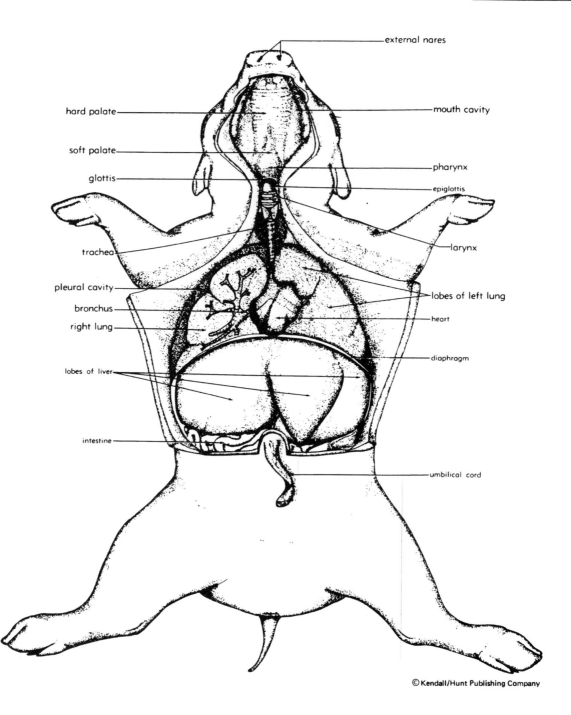

**Figure 23.4** Fetal Pig, Respiratory System

## Part IV: Structures of the Neck Region and Mouth

Using a blunt probe, carefully separate the structures in the neck region and observe:

1. The **larynx**, a large, firm white oval structure near the anterior portion of your incision.
2. The **trachea**, a tube leading from the larynx into the lungs. The trachea can be distinguished by cartilage rings which keep it from collapsing.
3. The **thyroid gland**, a small reddish ball-like structure lying on the trachea slightly posterior to the larynx.

Name _____    Section _____

4. The **thymus gland**, a very large whitish globular structure running down each side of the neck, with a portion of it covering part of the pericardium. This gland is in close physical contact with the neck muscles.
5. The **esophagus**, a tube just dorsal to the trachea. The esophagus carries food from the pharynx to the stomach.

Using a scissors, cut through the muscles on each side of the cheek by putting one point of the scissors in the pig's mouth so that the lower jaw can be pulled posteriorly. The bones of the lower jaw may need to be pulled or cut from their attachment to the skull. The lower jaw should flop down onto the pig's chest. Observe:

6. The **hard palate**, the washboardlike "roof" of the mouth.
7. The **soft palate**, posterior to the hard palate.
8. The flaplike **epiglottis** which can fold back to cover the opening of the larynx during swallowing.
9. The **glottis** which is the opening of the larynx.
10. From the mouth, insert a blunt probe down the **trachea** and later down the **esophagus**. Then feel for the tip of probe in the neck region.

**Part V: Circulatory System**

Refer to figures 23.5, 23.6 and 23.7 and locate the blood vessels listed below.

**ARTERIES**—carry blood away from the heart.

1. **Dorsal aorta:** Large artery located just ventral to the vertebral column.
2. **Aortic arch:** The aorta leaves the left ventricle anteriorly then arches becoming the dorsal aorta.
3. **Pulmonary artery:** Leaves the right ventricle and goes to the lungs. Most prominent vessel on top of the heart.

    *Note:* As you examine the fetal pig, the aortic arch is "behind" the pulmonary artery.

4. **Brachiocephalic or innominate artery:** First, very short branch off the aortic arch. Brachiocephalic artery immediately branches into the right subclavian artery and the bicarotid trunk which branches into the right and left common carotid arteries.
5. **Right subclavian artery:** Runs into the right shoulder.
6. **Left subclavian artery:** Second branch off the aortic arch; runs into the left shoulder.
7. **Right and left common carotid arteries:** Branches off the bicarotid trunk and goes into the head. Common carotids will branch to form internal and external carotid arteries.
8. **Right and left renal arteries:** Branches off the dorsal aorta running to the kidneys.
9. **Right and left genital arteries:** Small vessels branching from the dorsal aorta just posterior to the renal arteries; run to testes or ovaries.
10. **Right and left external iliac arteries:** Large branches from the dorsal aorta running into the back legs.
11. **Right and left umbilical arteries:** Large arteries lying on each side of the urinary bladder. The first part of the internal iliac arteries forms the connection between the dorsal aorta and the umbilical arteries.

**VEINS**—carry blood toward the heart.

12. **Anterior vena cava:** Also called **superior vena cava**, precava or precaval vein; short large blood vessel entering the right atrium bringing blood from the head; formed by joining of innominate veins.
13. **Posterior vena cava:** Also called **inferior vena cava**, post cava or postcaval vein; blood vessel entering the right atrium bring blood from the posterior regions.
14. **Right and left subclavian veins:** Branch from the shoulders entering the innominate veins.
15. **Right and left external jugular veins:** Larger, more lateral of the two pair of jugular veins.

Name _____  Section _____

16. **Right and left renal veins:** Run from kidneys to the posterior vena cava; near the renal arteries.
17. **Right and left genital veins:** Small vessels running from testes or ovaries to the posterior vena cava.
18. **Right and left common iliac veins:** Large veins carrying blood from the back legs to the posterior vena cava.
19. **Umbilical vein:** Small vein returning blood to the fetal circulatory from the placenta. It was cut and tied with string during initial dissection.

**HEART—**
20. **Right atrium:** Also called right auricle; saclike right anterior portion of the heart receiving the large anterior and posterior vena cavae. The right atrium appears to sit on top of the heart.
21. **Right ventricle:** Large muscular right posterior portion of the heart below the right atrium.
22. **Left atrium:** Also called left auricle; saclike left anterior portion of the heart.
23. **Left ventricle:** Larger muscular left posterior portion of the heart below the left atrium.
24. **Pulmonary artery:** Large artery leaving the right ventricle, carries non-oxygenated blood to the lungs.
25. **Aorta:** Large artery leaving the left ventricle.

*The following structures need to be located on the heart model only.*

26. **Tricuspid valve:** Three-flapped valve between right atrium and right ventricle.
27. **Bicuspid valve:** Also called **mitral valve**; two-flapped valve between left atrium and left ventricle.
28. **Chordae tendineae:** Cords attaching tricuspid and bicuspid valves to their ventricles so the valves are not pushed into the atria by the pressure of blood at contraction.
29. **Papillary muscles:** attach chordae tendineae to the ventricles.
30. **Pulmonary veins:** Four small veins entering the left atrium from the lungs carrying oxygenated blood.
31. **Semilunar valves:** Half-moon shaped valves in the pulmonary artery and aorta preventing backflow of blood into the heart.
32. **Coronary arteries:** Arteries lying on the surface of the heart supplying blood to the heart muscle.

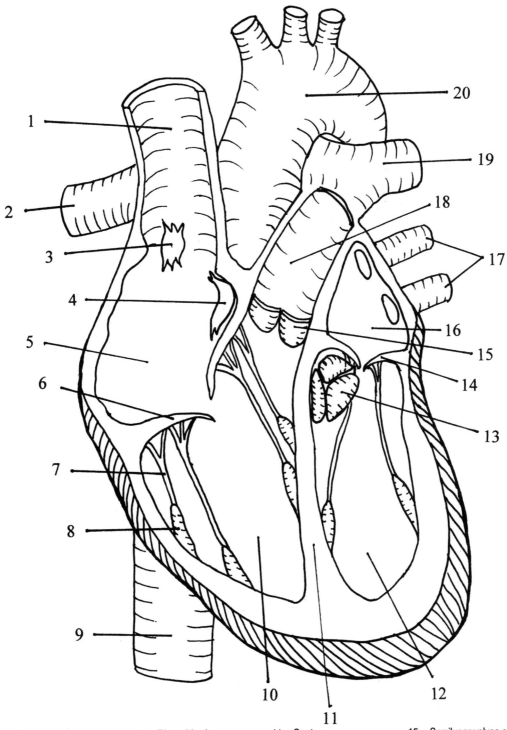

1. Anterior vena cava
2. Right pulmonary artery
3. Sinoatrial node
4. Atrioventricular node
5. Right atrium
6. Tricuspid valve
7. Chordae tendineae
8. Papillary muscle
9. Posterior vena cava
10. Right ventricle
11. Septum
12. Left ventricle
13. Semilunar valves of aorta
14. Bicuspid valve
15. Semilunar valves of pulmonary
16. Left atrium
17. Pulmonary veins
18. Pulmonary artery
19. Left pulmonary artery
20. Aorta

**Figure 23.5** Human Heart

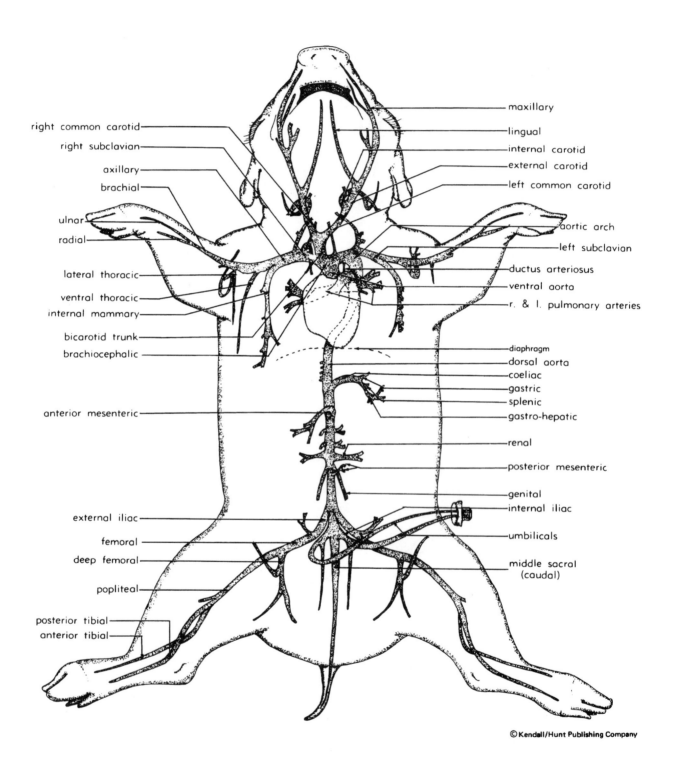

**Figure 23.6** Fetal Pig, Arterial System

206

Name _____  Section _____

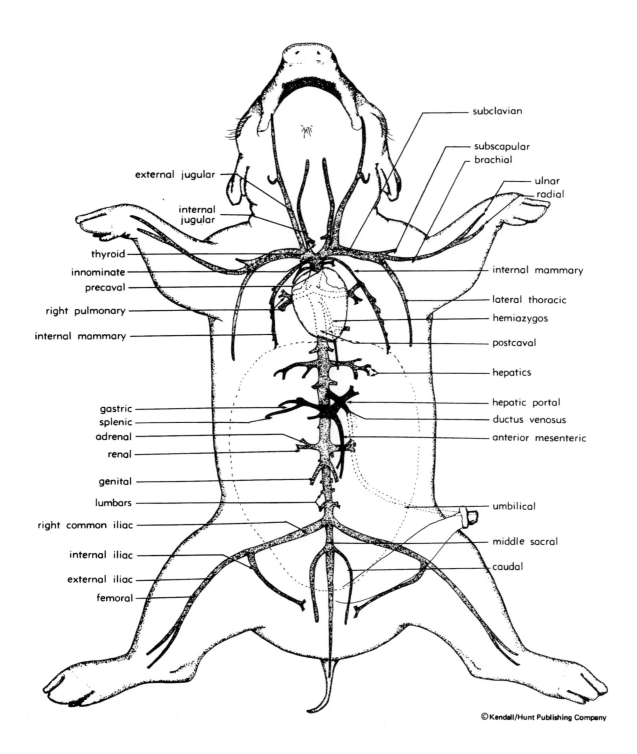

**Figure 23.7** Fetal Pig, Venous System

Name _____  Section _____

## Part VI: Reproductive System

You will be dissecting a specimen of one sex therefore you will need to work with another group having a pig of the opposite sex.

Refer to Figures 23.8 through 23.11 and locate the structures below.

### Female Reproductive System

1. The **ovary** appears as a small whitish oval structure on the dorsal side of the abdominal cavity below the kidney. There is one ovary on each side.
2. An **oviduct** or **Fallopian tube** or uterine tube is a tiny twisted tube coiled around the surface of the ovary then going to the middle of the body where it connects to the uterus.
3. The **uterus** is located at the medial ends of the oviducts. The uterus runs dorsal-caudal to the urinary bladder.
   Pull the urinary bladder forward. Use a scissors to cut through the body wall exactly on the midline on the ventral side of the pelvic bones so they can be spread apart. Severely stretch the pig by pulling sideways on each hind leg to help expose some of the structures associated with the reproductive system.
4. The **urethra** runs out from the **urinary bladder** and joins the **urogenital sinus** which runs out of the body. Humans do not have a urogenital sinus. In humans, the urethra exits directly to the outside.
5. The **urogenital papilla** is a small fleshy structure protruding externally, ventral to the tail in pigs. This structure is found in pigs but not in humans.
6. The **vagina** is located dorsal to the urethra. The vagina and urethra connect to the urogenital sinus. At its other end, the vagina connects to the uterus.
7. The **cervix**, which is the lower part of the uterus, can be distinguished as a firm structure between the vagina and the body of the uterus.

### Additional Structures on the Human

8. The **labia majora** are fleshy and more prominent folds of skin located externally to the small folds of skin, the **labia minora**.
9. The **clitoris**, a homolog to the penis, is located at the anterior junction of the labia majora.
10. The **endometrium** is the lining of the uterus that builds then deteriorates on a monthly cycle.
11. The **myometrium** is the smooth muscle portion of the uterus outside the endometrium.
12. The **ovarian (genital) artery** is a small artery branching off the dorsal aorta to the ovary.
13. The **umbilical cord** runs between the **placenta** and the **fetus**.

Name _____    Section _____

## Male Reproductive System

To dissect the reproductive system of a male pig, use a scissors to cut through the body wall just *off the midline*, beginning near the umbilical cord and proceeding posteriorly. Cut through the pelvic bone. Severely stretch the pig by pulling sideways on each hind leg.

1. The **penile opening** is the small opening of the male urogenital system located just caudal to the umbilical cord. The penile opening marks the end of the urethra.
2. The **penis** is located just under the skin from the penile opening toward the tail region. The penis is internal in the pig and poorly developed at this age.
3. The **urethra** runs through the penis from the penile opening to the Cowper's glands, which are located at a bend in the urethra dorsal to the pelvic bone. Carefully dissect out the urethra along its entire length. It will be necessary to cut through some muscle and bone to locate the urethra as it leaves the **urinary bladder**.
4. The **testes** are paired, oval structures located in the scrotal sac, which has been cut open during the dissection of the urethra. Locate one of the testes. Dissect off the tough connective tissue covering the testes.
5. The **epididymis** is a tightly coiled tubule lying on the surface of the testis. The epididymis connects to the vas deferens.
6. The **vas deferens** or **ductus deferens** is the tube running through the inguinal canal out of each scrotal sac. The vas deferens from each testis can be seen inside the abdominal cavity where they empty into the urethra by joining together to form the **ejaculatory duct.**
7. The **seminal vesicles** are paired glands located dorsally where the **ejaculatory duct** forms. The **prostate gland** is small in fetal pigs and difficult to see. The prostate gland is located between the seminal vesicles.
8. The **Cowper's glands** or **bulbourethral glands** are large elongate whitish structures located on each side of the urethra at the bend in the urethra where it joins the penis.

## Additional Structures on the Human

9. The **scrotum** is the sac-like structure enclosing and suspending the testes and ducts outside the abdominal cavity.
10. The **prostate gland** is a gland located at the junction of the **ejaculatory duct** and the **urethra**.
11. The **corpus cavernosum** is erectile tissue running nearly the length of the penis. The corpus cavernosum is anterior to the corpus spongiosum.
12. The **corpus spongiosum** is the erectile structure located around the urethra in the penis.
13. The **seminiferous tubules** are highly coiled tubules located in each lobule of the testis. Sperm develop in these tubules.
14. The **spermatic (genital) artery** is a small artery branching off the dorsal aorta to the testis.

Name _____  Section _____

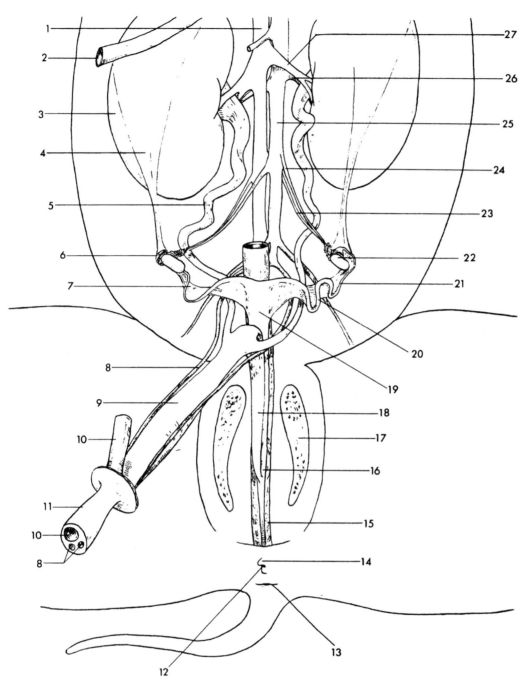

1. Post caval vein
2. Duodenum
3. Kidney
4. Mesovarium
5. Ureter
6. Funnel
7. Broad ligament
8. Umbilical artery
9. Urinary bladder
10. Umbilical vein
11. Umbilical cord
12. Urogenital orifice
13. Anus
14. Genital papilla
15. Rectum
16. Vagina
17. Pubic symphysis (separated)
18. Urethra
19. Uterus
20. Uterine horn
21. Fallopian tube
22. Ovary
23. Ovarian vein
24. Ovarian artery
25. Aorta
26. Renal artery
27. Renal vein

Illustration by Carolina Biological Supply Company, © 1972

**Figure 23.8** Fetal Pig, Female Urogenital System

Name _____   Section _____

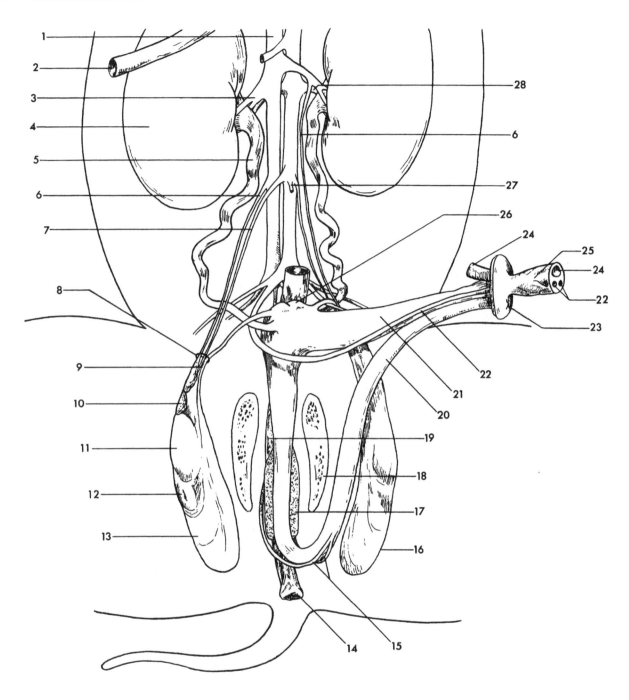

1. Post caval vein
2. Duodenum
3. Renal vein
4. Kidney
5. Ureter
6. Spermatic vein
7. Spermatic artery
8. Spermatic cord
9. Vas deferens
10. Head of epididymis
11. Testis
12. Tail of epididymis
13. Gubernaculum
14. Anus
15. Retractor penis muscle
16. Scrotal sac
17. Bulbo-urethral gland
18. Pubic symphysis (separated)
19. Rectum
20. Penis
21. Urinary bladder
22. Umbilical artery
23. External urethral orifice
24. Umbilical vein
25. Umbilical cord
26. Seminal vesicle
27. Posterior mesenteric artery
28. Renal artery

Illustration by Carolina Biological Supply Company, © 1972

**Figure 23.9** Fetal Pig, Male Urogenital System

Name _____   Section _____

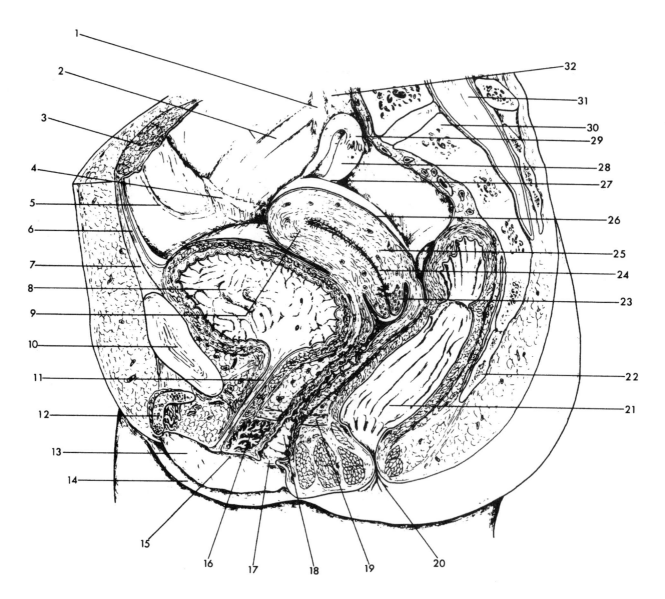

1. Suspensory ligament, ovary
2. External iliac a. and v.
3. Rectus abdominis
4. Round ligament
5. Lateral umbilical ligament
6. Middle umbilical ligament
7. Rectus ligament
8. Urinary bladder
9. Uterus
10. Pubic symphysis
11. Urethra
12. Clitoris
13. Labium minus
14. Labium majus
15. Paraurethral duct
16. Paraurethral gland
17. Vaginal orifice
18. Vagina
19. Urogenital diaphragm
20. Anus
21. Rectum
22. Coccyx
23. Cervix
24. Tunica mucosa
25. Tunica muscularis
26. Tunica serosa
27. Ovarian ligament
28. Ovary
29. Uterine tube
30. Sacrum
31. Spinal cord
32. Ureter

Illustration by Carolina Biological Supply Company, © 1972

**Figure 23.10** Human Female Pelvis

Name _____   Section _____

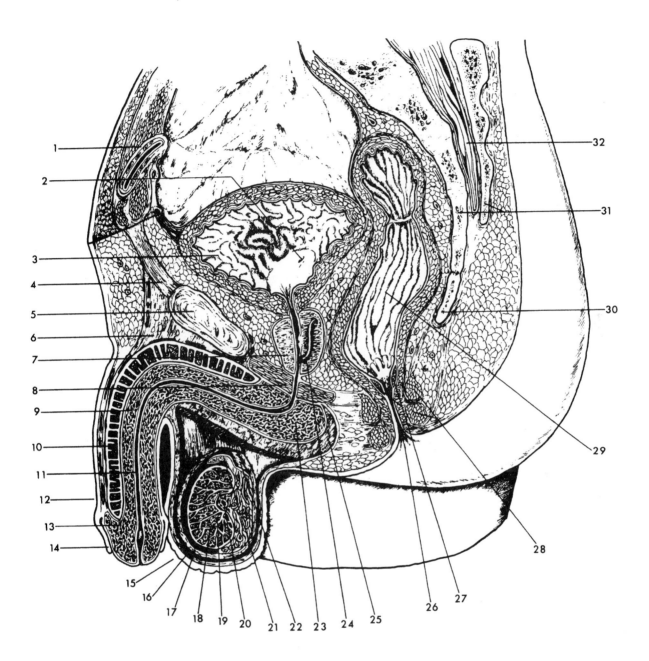

1. Inguinal ligament
2. Orifice of ureter
3. Urinary bladder
4. Rectus ligament
5. Pubic symphysis
6. Suspensory ligament of penis
7. Prostate gland
8. Urogenital diaphragm
9. Corpus spongiosum
10. Corpus cavernosum
11. Urethra
12. Penis
13. Glans
14. Foreskin
15. Scrotum
16. Dartos m.
17. Cremaster m.
18. Septum
19. Lobule
20. Rete testis
21. Epididymis
22. Vas deferens
23. Bulbus penis
24. Ejaculatory duct
25. Bulbocavernosus m.
26. Anus
27. Internal anal sphincter m.
28. External anal sphincter m.
29. Rectum
30. Coccyx
31. Sacrum
32. Spinal cord, conus

Illustration by Carolina Biological Supply Company, © 1972

**Figure 23.11** Human Male Pelvis

Name _____   Section _____

## Part VII: Functions of the Major Structures

Use your textbook and reference books in the library to find the function(s) of the following structures.

Diaphragm: _____

Liver: _____

Gallbladder: _____

Common bile duct: _____

Esophagus: _____

Stomach: _____

Small intestine: _____

Large intestine: _____

Urinary bladder: _____

Kidneys: _____

Ureter: _____

Larynx: _____

Trachea: _____

Lungs: _____

Epiglottis: _____

Pancreas: _____

Thymus gland: _____

Thyroid gland: _____

Mesentery: _____

Pericardium: _____

Name _____ Section _____

Right atrium: _____

Right ventricle: _____

Left atrium: _____

Left ventricle: _____

Dorsal aorta: _____

Carotid arteries: _____

Iliac arteries: _____

Anterior vena cava: _____

Posterior vena cava: _____

Chordae tendineae: _____

Ovary: _____

Uterus: _____

Clitoris: _____

Testes: _____

Epididymis: _____

Prostate gland: _____

Seminal vesicle: _____

Cowper's gland: _____

Corpus cavernosum/corpus spongiosum: _____

Seminiferous tubules: _____

Name _____   Section _____

# LAB 24

# Feeding Activity in Small Animals

**Problem**

How do Hydra and Planaria differ in their feeding techniques and response to stimuli?

**Objectives**

After completing this lab, the student will be able to:

1. Describe feeding techniques in two animals.

2. Describe how the two animals respond to a variety of stimuli.

3. Contrast the muscle activity of a sessile animal to a free-living animal.

**Preliminary Information**

This lab will explore the feeding mechanisms used by a fresh-water cnidarian with radial symmetry and a platyhelminthes or flatworm with bilateral symmetry. The cnidarian, *Hydra*, spends its adult life attached to the substrate. Its radial symmetry is an advantage because sessile animals such as the *Hydra* can take in food coming from any direction. The flatworms or *Planaria*, however, actively seek food. For animals moving around seeking food, it is best to have the sense organs concentrated in the part of the body that first encounters the environment. This, of course, is the head. The concentration of sense organs in the head area is called **cephalization**. Simpler animals than *Planaria*, like sponges and sea anemones and coral, lack cephalization. It is a survival advantage to be aware of the immediate environment so cephalization with bilateral symmetry is the dominant arrangement seen in the animal kingdom.

The tentacles of the *Hydra* are covered with nematocysts or stinging cells. When anything touches the tentacles it causes the stinging cells to fire. The injected poison from these stinging cells will immobilize the prey giving the *Hydra* ample opportunity to ingest the prey.

The digestive system of the *Hydra is* a simple sac called a gastrovascular cavity with a mouth on one end. There is not a second opening for waste so any indigestible materials are expelled back through the mouth. This isn't a particularly sophisticated digestive system. The *Daphnia*, the food organism in this lab, is a slow moving fresh-water crustacean that is generally easy for the *Hydra* to catch.

As you watch a *Planaria* move around the dish, notice that it lacks any special breathing apparatus. This animal also lacks blood. Animals that seek food use more energy looking for food than the *Hydra*. When an organism requires more energy, more food and oxygen must be consumed to release extra energy. Yet *Planaria* lack specialized respiratory structures. Most flatworms solve this dilemma by dropping out of the "rat race" by not competing and becoming parasitic. However, *Planaria* are nonparasitic and free-living. They can be found under rocks in clean streams. *Planaria* have a definite head. The mouth, not located in the head region, is at the end of a tube or proboscis coming from mid-ventral part of the flatworm. So

Name _____  Section _____

look for a tube on the bottom side of the flatworm located half way back. *Planaria*, like *Hydra*, possess a gastrovascular cavity that digests and distributes the food.

## Part I: Hydra

1. Set up a binocular microscope at your lab table.
2. Obtain a *Hydra* in a syracuse dish with *pond water* from the supply area. Make sure the *Hydra* has been in the dish for at least 5 minutes. Dry the bottom of the dish before placing it on the binocular microscope.
3. Observe the *Hydra* using both magnifications. NOTE: on the right side of the main body of the scope is a knob that rotates between 15X and 30X. Use the magnification that shows the animal the best.
4. At this point, review the data and discussion parts so you have a clear idea what you need to observe. Sketch the *Hydra* and record your observations.
5. Cover one half of the syracuse dish with foil. Wait for five minutes and record data.
6. Using a wide tip dropping pipette, add a single *Daphnia* or other food in the area close to the *Hydra*. Record your observations.
7. If the *Hydra* is unable to catch the *Daphnia*, use a fine tipped forceps to feed the *Daphnia* or other food to the Hydra.

## Part II: Planaria

1. Use a paint brush to remove one planarian from the enamel tray in the supply area. Place the planarian in a syracuse dish filled with *pond water*. Dry the bottom of the dish before placing it under the binocular scope.
2. For your observations, turn both lights on and observe if the behavior changes. Sketch the planarian and record your observations.
3. Use both ends of a paint brush and touch planarian in several places. Record the data.
4. Use the brush as barrier to forward motion. Record the data.
5. Cover one half of the syracuse dish with foil. Wait for five minutes and record data.
6. Shut off both microscope lights, cover the dish with foil for five minutes. Remove the cover but **leave the lights off**. Put a drop of blood from your finger or beef liver close to the flatworm. Record your observation for five-ten minutes.

Name _____  Section _____

## Part II: Data

| Sketch a nonfeeding *Hydra* | Sketch a feeding *Hydra* |
|---|---|
| | |

| Sketch a nonfeeding planarian | Sketch a feeding planarian |
|---|---|
| | |

Name _____ Section _____

## Hydra Data

1. Describe the *Hydra* behavior before the food is added. Include the following information.
   a. Is the *Hydra* attached to the dish?




   b. Is the body extending, contracting or both?




   c. What are the tentacles doing?




   d. Does the *Hydra* show any preference for light or dark? If so, which?




2. After the *Daphnia is* added, record your observations for 10 minutes.
   a. Is the *Hydra* attached to the dish?




   b. Is the body extending, contracting or both?




   c. What are the tentacles doing?




   d. Other observations.

Name _____   Section _____

3. Using a paint brush, touch the *Hydra* in several places with both ends of the brush. How does the *Hydra* respond to touch?

## Planarian Data

1. Record your observations of Planaria under the following conditions:
   a. Microscope with no lights:

   b. Microscope with lights:

   c. Touch:

   d. Barrier:

   e. Cover one half of dish:

   f. Feeding:

Name _____  Section _____

## Part IV: Discussion

1. Compare the touch response of the *Hydra* to the planarian.




2. Which animal probably has the easier time finding food? Why?




3. Which animal responds to stimuli the best? How can you explain this?




4. What advantage does radial symmetry have for the *Hydra?*




5. What advantage does bilateral symmetry have for the planarian?

Name _____          Section _____

# LAB 25

# Kidney Function and Urinalysis

**Developed by Sandra Gibbons**

**Problem**

What can be learned about a person's health and kidney function by analyzing their urine?

**Objectives**

After completing this lab exercise, the student will be able to:

1. Explain the stages of urine production.
2. Use a Multistix 9 SG reagent strip to determine the composition of a urine sample.
3. Discuss which urinalysis tests are abnormal and the probable cause behind the abnormality.
4. Identify the parts of a nephron.

**Preliminary Information**

The kidney is an important organ in the excretory system. It is responsible for eliminating toxic nitrogenous wastes produced during metabolism as well as controlling the water and salt concentration in the blood. The functional unit of the kidney is the nephron, which is actually a single long tubule associated with many blood vessels. Each kidney contains thousands of nephron tubules.

There are several stages to the production of urine by the nephron. The first stage is known as **filtration**. A cup-shaped swelling of the nephron known as the **glomerular capsule** surrounds a ball of capillaries known as the **glomerulus**. The glomerulus is much more permeable than other capillaries. The blood pressure in this region forces all small molecules from the blood and into the glomerular capsule. Filtration removes substances from the blood based on size. Small molecules such as water, glucose, amino acids, salts, urea, uric acid and creatinine are forced out of the blood at this point. Large molecules such as proteins and blood cells remain in the blood.

Since many of the molecules removed during the filtration step are necessary for life (water, glucose, amino acids) they need to be taken out of the nephron filtrate and put back into the blood. This happens during the **selective reabsorption** stage which occurs along the **proximal convoluted tubule** of the nephron. During this stage molecules such as water, glucose, amino acids, and salts move from the proximal tubule of the nephron back into the capillary network of the blood.

The next region of the nephron is known as the **loop of the nephron**. Along this portion water and salt will be moved out of the filtrate and back into the blood. **Tubular secretion** occurs at the portion of the nephron known as the distal convoluted tubule. Substances such as uric acid, creatinine, hydrogen ions, ammonia, and antibiotics such as penicillin are removed from the blood and actively transported into the nephron at this point. The fluid coming from the distal convoluted tubule will enter a **collecting duct** which will collect the fluid (urine) produced by many nephrons and carry it to the renal medulla portion of the kidney from where it will pass into the ureter going to the bladder and then to the urethra and finally out of the body.

The chemical analysis of urine (urinalysis) can tell the state of health of an individual. During this laboratory you will be testing five artificial urine samples using Multistix 9 SG reagent strips which will simultaneously test the glucose, bilirubin, ketone, specific gravity, blood, pH, protein, nitrite, and leukocyte level in the urine. Table 24-2 lists some causes for the various levels you may observe.

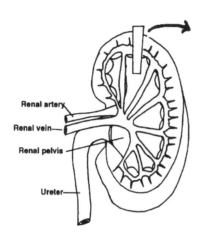
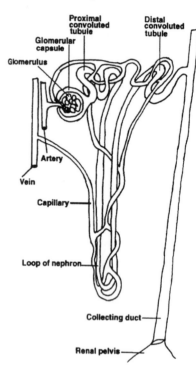

**Figure 25.1** The Nephron

## Part I: Methods and Materials

Work as a lab table.

**Note**: Wash all test tubes in soapy water and rinse before and after using. Make a label on each test tube by using a grease pencil or masking tape wrapped completely around the top, then label with a permanent marker.

1. Obtain 5 test tubes from your glassware drawer and label them A, B, C, D, and E.

2. From the stock table area, obtain the appropriate artificial urine sample (A-E) and fill each test tube approximately half way with the correct sample.

3. Obtain a bottle of Multistix 9 SG reagent strips from the stock table. You need to familiarize yourself with the strips before you test any urine. The strip contains 9 colored squares, each of which tests for a different substance in the urine. The label on the Multistix 9 SG container shows you which test each square represents as well as when you should read your results.

Name _____   Section _____

4. You will use a different reagent strip to test each of the urine samples. To conduct the test dip the reagent strip entirely into the urine sample to insure that all squares are in contact with the urine. Remove the strip immediately and start timing. When 30 seconds have elapsed read the result for the glucose and bilirubin test; when a total of 40 seconds have elapsed read the ketone test; at 45 seconds read the specific gravity test; at 60 seconds read the blood, pH, protein, and nitrite tests. At 2 minutes read the leukocyte test. Record the data in the table.

5. Place the used reagent strip on paper towel.

6. Optional: You may conduct these tests on your own urine.

7. Discard the reagent strips in the garbage, discard the artificial urine samples in the sink and wash, rinse, and dry the test tubes.

**Part II: Data**

TABLE 25.1 Results of Urinalysis on Each Urine Sample A-E

| Test | Samples | | | | | |
|---|---|---|---|---|---|---|
| | A | B | C | D | E | Student |
| glucose | | | | | | |
| bilirubin | | | | | | |
| ketone | | | | | | |
| specific gravity | | | | | | |
| blood | | | | | | |
| pH | | | | | | |
| protein | | | | | | |
| nitrite | | | | | | |
| leukocyte | | | | | | |

Name _____   Section _____

**TABLE 25.2** Possible Causes for Abnormal Test Results

| Test Result | Medical Cause | Other Cause |
|---|---|---|
| glucose present | uncontrolled diabetes | a large meal eaten recently |
| bilirubin present | bile pigment breakdown | high levels indicate liver problems |
| ketone present | uncontrolled diabetes | produced during fat metabolism, could be due to dieting |
| specific gravity | | |
|   low <1.010 | severe kidney damage | increased fluid intake |
|   high >1.025 | uncontrolled diabetes or severe anemia | decreased fluid intake or loss of fluid |
| blood | infection | menstruation |
| pH | | |
|   low <4.5 | uncontrolled diabetes | high protein diet or cranberry juice in diet |
|   high >8.0 | severe anemia | diet rich in vegetables or dairy products |
| protein | severe anemia or infection | high protein diet |
| nitrite | infection | none |
| leukocyte | infection | none |

## Part III: Discussion and Conclusions

1. Discuss the results of each sample including the probable medical problem associated with each sample.

   A

   B

   C

   D

   E

Name _____ Section _____

2. Based on what you know about the functioning of the nephron, at which point would glucose be entering into the filtrate (urine)?

3. At which point along the nephron would an antibiotic enter the filtrate (urine)?

4. You are being treated with antibiotics for a urinary tract infection. How could you use the Multistix to see if the antibiotics are working?

Name _____   Section _____

# LAB 26

# Chemical Effects on the Respiratory Rate of Brine Shrimp

## Problem

What effect does epinephrine and acetylcholine have on the respiratory movements of brine shrimp?

## Objective

After completing this lab, the student will be able to:

1. Identify the effects of epinephrine and acetylcholine on the respiration rate of brine shrimp.

## Preliminary Information

Brine shrimp are crustaceans commonly used as fish food as they are easy to raise and their eggs are plentiful. Unlike many of the larger crustaceans, the brine shrimp normally swim with their ventral side up. The feathery gills of the brine shrimp are located on the anterior appendages. The brine shrimp carries on gas exchange by waving its legs through the water. The number of movements per minute can be used as an indication of its respiratory rate.

## Part I: Method and Materials

1. Obtain a microscope slide with a depression in the center.
2. Using a spoon, add one brine shrimp and water to the slide.
3. If the slide begins to dry out at any time during the experiment, add water from the original brine shrimp container.
4. Using a binocular dissecting scope and a hand counter, count and record the number of leg movements per minute for three minutes. This is the first control count.
5. Use a paper towel to draw off some of the water on the brine shrimp. CAUTION: Don't allow the slide to dry out.
6. Add one drop of epinephrine. Wait one minute then count and record the number of leg movements per minute for three minutes.
7. Use a paper towel to draw off the excess fluid.
8. Flood the animal with water from its original container. Wait five minutes.
9. Flood the animal with water from its original container a second time.
10. Using a dissecting scope and a hand counter, count and record the number of leg movements per minute for three minutes. This is the second control count.
11. Use a paper towel to draw off the excess fluid.
12. Add two drops of acetylcholine. Wait three minutes then count and record the number of leg movements per minute for three minutes.

Name _____  Section _____

13. Put the brine shrimp in the container labeled used brine shrimp.
14. Record the data in the table.

## Part II: Data

**TABLE 26.1**
Respiratory Movements of Brine Shrimp

| Time | First Control | Epinephrine | Second Control | Acetylcholine |
|---|---|---|---|---|
| First Minute | | | | |
| Second Minute | | | | |
| Third Minute | | | | |
| Average | | | | |
| Percent Change* | | | | |

*Percent Change = $\dfrac{\text{Average Experimental - Average Control}}{\text{Average Control}} \times 100$

When calculating percent change for epinephrine, use first control count.

When calculating percent change for acetylcholine, use second control count.

## Part III: Discussion

1. Compare the percent change of epinephrine to the percent change of acetylcholine.

2. Discuss the organisms response to epinephrine and acetylcholine. What effect does epinephrine have? What effect does acetylcholine have?

Name _____  Section _____

# LAB 27

# Chemicals Effecting Muscle Contraction

## Problem

What cellular structures and chemical substances are necessary for contraction?

## Objectives

After completing this lab, the student will be able to:

1. List the differences between smooth, skeletal, and cardiac muscle, including microscopic appearance.
2. Identify the substance which directly supplies the energy for muscular contraction.
3. Name the two chemicals that must be present for proper muscle contraction.

## Preliminary Information

Muscles are the primary effectors in biological systems. Upon activation, muscles convert stored chemical energy into mechanical work. There are three types of muscles: Skeletal or striated, smooth, and cardiac. Muscle contraction is accomplished by two primary muscle proteins, **actin** and **myosin**, sliding past one another. Once contracted, a muscle does not have the ability to expand itself. It must be pulled back to its original length. This is accomplished by an **antagonistic muscle**. For example, the biceps and triceps muscles in the human arm are antagonistic muscles. Together muscles and nerves are responsible for reflexes and other more complex behavioral patterns.

The process of muscle contraction requires several factors. One factor is the positive metal ion of a salt which "turns on" the process of muscle contraction by combining with one of the muscle proteins. Another factor is the energy source, **ATP**, which combines with another muscle protein. When the ATP splits, the released energy is used to effect the ratch-like movement of other muscle proteins past one another. This sliding of muscle proteins past one another results in the shortening of the entire muscle, thus the process of muscle contraction.

## Part I: Types of Muscle Tissues

1. Obtain three prepared microscope slides:
   a. striated (skeletal) muscle [This slide contains both longitudinal section (labeled 'l.s.') and cross section (labeled 'x.s' or 'c.s.')],
   b. cardiac muscle,
   c. smooth muscle.

Cardiac muscle and smooth muscle are in longitudinal section.

Name _____  Section _____

2. Locate a few cells on each slide. Under high power, diagram and label them on the data sheet. Note the following:
   a. striated muscle, l.s.—These cells are about 2 to 3 cm long with cross striations and many nuclei along the side of each cell.
      Label: **muscle cell**, **striation**, **nucleus**.

   b. striated muscle, x.s.—In the end view, notice the round appearance of the cells and the arrangement of the internal proteins.
      Label: **muscle cell**, **proteins**.

   c. cardiac muscle, l.s.—These cells interlock with each other on their ends forming intercalated discs which can be seen as darker lines than the straiations. There is one nucleus per cell.
      Label: **intercalated disc**, **striation**, **nucleus**.

   d. smooth muscle, l.s.—These are no-striated, long, thin, tapered, thread-like cells with a nucleus near the middle of each cell.
      label: **nucleus**.

Name _____   Section _____

## Part II: Source of Energy for Muscle Contraction

One result of cell respiration is the capture of energy from chemical bonds in the form of ATP. The energy in ATP is able to drive cellular metabolic activities. Muscle contraction, a common energy-consuming activity, and the substances necessary for contraction to occur will be studied in this exercise.

The purpose of this exercise is to determine what the proper substances are for muscle contraction.

1. Wash three slides in soapy water, rinse and dry.
2. On a *clean* microscope slide, obtain two very thin muscle fibers which can be torn off of a commercially prepared psoas muscle of a rabbit using a dissecting microscope and fine forceps. Place the fibers in a minimal amount of glycerol-water mixture to prevent drying. Position the fibers straight and parallel to each other.
3. Determine the lengths of the strands by measuring them with a millimeter ruler.
4. Record the lengths in the table in the data section.
5. Obtain a small bottle of **ATP plus salt** (potassium and magnesium ions) solution from the supply area. **Note: Do not touch tip of dropper to muscle.** Carefully flood the slide with several drops of this solution without splashing. Observe the reaction of the fibers. **Immediately return the bottle to the supply area.**
6. After 30 seconds or more, without touching the strands, remeasure the fibers and record the final length. Calculate the percent of contraction (formula is provided).
7. Repeat the experiment a second time using a clean slide, new fibers, and a solution of **ATP alone**.
8. Repeat the experiment a third time using a clean slide, new fibers, and a solution of **salt alone**.

Name _____  Section _____

## Part III: Data

**TABLE 27.1**
Types of Muscle Tissue

| Striated (Skeltal) Muscle—longitudinal section | Striated (Skeltal) Muscle—cross section |
|---|---|
| | |

| Cardiac Muscle | Smooth Muscle |
|---|---|
| | |

Name _____  Section _____

## TABLE 27.2
Muscle Contraction

| Slide 1 | Initial Length | Final Length After Adding ATP and Salt Solution | Percent Contraction* | Average Percent Contraction |
|---|---|---|---|---|
| Fiber 1<br>Fiber 2 | | | | |
| **Slide 2** | **Initial Length** | **Final Length After Adding ATP Only** | **Percent Contraction*** | **Average Percent Contraction** |
| Fiber 1<br>Fiber 2 | | | | |
| **Slide 3** | **Initial Length** | **Final Length After Adding Salt Only** | **Percent Contraction*** | **Average Percent Contraction** |
| Fiber 1<br>Fiber 2 | | | | |

*Percent contraction can be calculated by using this formula:

$$\text{Percent concentration} = \frac{\text{Initial length} - \text{Final length}}{\text{Initial length}} \times 100$$

## Part IV: Discussion

1. Which types of muscle cells are most similar in appearance? Provide evidence to support your statement.

2. If a muscle can only shorten itself, how can an arm be returned to its starting position?

3. What chemical factors must be present to cause muscle contraction?

Name _____    Section _____

# LAB 28

# Human Reflexes and Senses

## Problem

Using a variety of experiments, can you determine if the senses are always reliable?

## Objectives

After completing this lab, the student will be able to:

1. Identify different reflex responses.
2. Discuss sensitivity to touch in different parts of the body.
3. Explain the effects of continuous exposure of a substance on the sense of smell and its subsequent reliability.

## Preliminary Information

As we touch, smell, hear and see things, we experience our environment. A loss of any one of the senses can be a major problem. People can adjust, of course, but the world is never the same.

Reflexes, like senses, are an important survival mechanism. They enable your body to respond to environmental changes quickly without your brain having to think about it. After the event has happened, such as pulling your hand off a hot stove, the brain becomes aware of the event. The ability to respond quickly can protect your body.

In this lab, you will explore some of the human reflexes and senses. Consider what would happen if these were no longer present.

## Part I: Touch Sensitivity

Working in pairs, determine areas of greater sensitivity by having the subject note when two pins can be felt as only one.

1. Obtain two dissecting pins and clean them with alcohol.
2. The subject should close his eyes while the experimenter tests the areas listed in the following table.
3. Each area is tested by the experimenter holding the pins 10 mm apart and very gently touching the subject with both pins at the same time. Move the pins closer together one mm at a time and touch the subject again. The subject should be told to indicate when both pins feel like one. Measure and record the distance between the pins in the table.
4. To interpret the data: the smaller the number, the more sensitive the area.

Name _____   Section _____

## TABLE 28.1
Sensitivity to Touch

| Area Tested | Distance Between Pins |
|---|---|
| Index Finger | |
| Back of the Hand | |
| Palm of the Hand | |
| Back of the Neck | |
| Inner Forearm Near Elbow | |
| Lips | |

It is important that the skin surface is just touched by the pins. What does that tell you about the relative depth of pressure receptors and pain receptors?

A tiny movement of skin hair can be sensed by nerve fibers. What is the advantage of this second tactile sense?

Which area tested is the most sensitive? Explain the advantage of having this area particularly sensitive?

**Part II: Ligament and Tendon Reflexes**

1. Patellar reflex. Working in pairs, have the subject sit on the middle of the table (sitting on the edge may break the table) so that the leg from the knee down hangs freely. The examiner should locate the patellar ligament (just below the knee cap or patella), then strike it with the flat edge of a reflex mallet. It is best for the subject to remain relaxed and divert his attention. Notice the degree to which the leg is extended by the contraction of the quadriceps muscle.

    Test the reflex on right and left legs. Do they respond the same?

2. Achilles jerk. Have the subject kneel on a chair with the feet hanging freely over the edge of the chair. Locate, then tap the tendon of Achilles in the heel of the foot. What is observed?

Name _____ Section _____

## Part III: Eye Reflexes

1. Corneal reflex. Quickly move the palm of your hand toward your partner's eyes, stopping a safe distance away. What is observed?

   What is the purpose of this reflex?

2. Photo-pupil reflex. Have the subject close his eyes for two minutes. Examine the pupil size immediately on opening the eyes. What is observed?

   What is the purpose of this reflex?

3. Convergence reflex. Have the subject look at a distant object (seven or more meters away). Note the position of the eyeballs. Have the subject immediately focus on a pencil held about 25 cm away. What change occurs in the position of the eyeballs?

   What is the purpose of this convergence?

## Part IV: Swallowing Reflex

1. Swallow the saliva in your mouth and immediately swallow again, then again. What do you observe?

2. The fact that swallowing can be performed in rapid succession is demonstrated by rapidly drinking a glass of water or some other fluid.

   Is swallowing a reflex? Explain.

Name _____  Section _____

**Part V: Olfactory Adaptation (Fatigue Time)**

1. Obtain a small bottle of isopropyl alcohol. Close one nostril and smell the solution. Exhale through your mouth. Continue doing this until you no longer smell the alcohol. Record the number of minutes before olfactory fatigue occurred.

2. Open the other nostril and smell the alcohol. What do you smell?

3. When olfactory fatigue occurs, are all olfactory cells affected or only the cells exposed to the stimulus?

4. Once you recovered and can smell the alcohol again, repeat the above procedure. When you have fatigued the olfactory nerves a second time, smell the bottle of oil of cloves. What do you smell? What conclusions can you make?

**Part VI: Sensations Travel at Different Speeds**

**A. Touch—Temperature Sensations**

Not all sensations move through the nervous system at the same speed. It has been hypothesized that sensations of highest survival value move most quickly.

Work in pairs for this experiment.

1. Obtain a metal spatula or blunt-tipped probe.
2. Hold it in your hand for a minute to warm it to body temperature.
3. Place the metal spatula on the top of your partner's bare foot.
4. Which sensation—touch or temperature—is sensed first? What is the approximate time delay between the sensations?

5. Place the spatula or probe in a small beaker of ice water for a minute. Repeat the experiment.

   Is there any difference in time between the sensations? If so, how does the speed of sensation of cold compare to the speed of sensation of warm?

Name _____    Section _____

## B. Response Time and Light Intensity

In this experiment you will measure time in bright light and dim light by measuring how quickly a meter stick can be grabbed.

1. Obtain a meter stick.
2. Your lab partner will catch the meter stick as it falls between his/her hands. Your partner should be standing with the palms of the hands 10 centimeters apart.
3. Hold the meter stick vertically with the zero mark even with the top of the hands.
4. Let go of the meter stick so it drops straight down. Your partner should catch it as quickly as possible.
5. Record the number where the meter stick was caught.
6. Repeat the experiment four more times, recording the data.
7. Trade places and repeat. Record the data.
8. Run the lights off and repeat. Record all data.

**Table 28.2**
Reaction Speed vs. Light

|         | Experiment #1 | | Experiment #2 | |
|---------|---------------|------|---------------|------|
|         | Light | Dark | Light | Dark |
| Trial 1 |       |      |       |      |
| Trial 2 |       |      |       |      |
| Trial 3 |       |      |       |      |
| Trial 4 |       |      |       |      |
| Trial 5 |       |      |       |      |
| Average |       |      |       |      |

9. How does light intensity affect reaction rate?

10. If you are driving a car at night, what adjustments should you make?

Name _____   Section _____

# LAB 29

# Control Systems: Experiment in Temperature Regulation

## Problem

What effect will changing the external environment temperature have on the regulation of body temperature?

## Objectives

Upon completion of this lab exercise the student will be able to:

1. Compare and contrast the similarities and differences between how an ectotherm and an endotherm regulate body temperature.
2. Define and give examples of ectothermic and endothermic animals.
3. Cite the major advantage endotherms possess over ectotherms.
4. Graph and tabulate the breathing response of an ectotherm and an endotherm when the temperature of its environment is changed.
5. Discuss the mechanisms, both internal and external, that ectotherms and endotherms utilize to regulate body temperature.

## Preliminary Information

**Homeostasis** is the maintenance of a steady internal state. Vertebrate animals attempt to maintain homeostasis in all aspects of their lives. When a vertebrate cannot maintain homeostasisis, its physiology is affected. For example, each animal has an optimum temperature range, and it will make adjustments (behavioral or physiological or both) to stay within this optimum range.

An **ectotherm** is an animal that obtains most of its body heat from the outside, rather than generating its own heat internally as an **endotherm** does.

When an ectotherm is exposed to an environment cooler than itself, it will lose heat to the environment and its body temperature and activity will decrease. Ectotherms modify their behavior to avoid exposure to temperature extremes.

Endotherms will also lose heat to the environment when the environment is cooler than the animal. However, the endotherm's metabolic rate will generate more heat to replace that which is being lost and the body temperature will remain constant.

Major sources of heat for organisms are the sun directly, environmental surroundings, metabolic reactions and muscle contractions. Endotherms are more efficient at utilizing these heat sources than ectotherms are.

Name _____    Section _____

The metabolic rate of an organism is the total sum of all chemical reactions and indicates how quickly or slowly these chemical reactions are taking place. The metabolic rate is directly related to cellular respiration of food molecules. If more food is broken down, more ATP energy is available for use and the metabolic rate increases. For the complete breakdown of food molecules, oxygen is necessary. Therefore, the amount of oxygen consumed is an indication of how fast the metabolic rate is functioning. In goldfish, who obtain their oxygen from gills, when more oxygen is needed, more water is moved across the gill membranes. So counting the movement of the operculum (gill cover) is an indication of oxygen consumption. By use of a volumeter type apparatus, the amount of oxygen consumed by a small mammal can be measured. Oxygen intake is considered to be an index of the rate of respiration or the rate of metabolism.

## Part I: Operculum Rates In Goldfish

### Material per two students

One goldfish in 150 ml or 250 ml beaker
One 400 ml or larger beaker
One stirring rod
One thermometer
Two beakers (for crushed ice and hot tap water)
Crushed ice.

### Method

CAUTION: The goldfish can go into shock if subjected to temperature changes that are too sudden and too abrupt. So when hot water or ice is added to the large beaker, do so gradually and stir frequently.

1. Add tap water until the larger beaker is half full and has a temperature of 22-26°C (room temperature).
2. Stir only with a stirring rod, NEVER use a thermometer.
3. Place the smaller beaker of aquarium water with the goldfish in the larger beaker, making sure the smaller beaker is stable and will not tip.
4. Watch the fish from above. Begin the experiment by counting the number of operculum (gill cover) movements per minute using a hand counter and record in the appropriate table. (See data sheets.)
5. Record the number of operculum movements each minute for three minutes.
6. Add ice to the outer beaker while monitoring the temperature of the inner beaker.
7. Decrease the temperature *gradually* by 2-3°C at a time until 14-18°C is reached.
8. Maintain the system at this temperature for **two minutes** before beginning the three minute count, record results.
9. Repeat the procedure until 4-8°C is reached, stabilize for two minutes and record results for three minutes.
10. *Slowly* warm the goldfish back up to 22-26°C by adding warm tap water to the outer beaker while monitoring the temperature in the inner beaker.
11. Repeat the count at 22-26°C to compare with initial run.
12. *Gradually* warm the goldfish to 32-36°C, stabilize and record count.

Name _____  Section _____

## Data

**TABLE 29.1**
Operculum Movements in Goldfish

| Temperature | Number Movements/Minute | Average |
|---|---|---|
| 22–26 °C  Exact Temp_____ | | |
| 14–18 °C  Exact Temp_____ | | |
| 4–8 °C  Exact Temp_____ | | |
| 22–26 °C repeat  Exact Temp_____ | | |
| 32–36 °C  Exact Temp_____ | | |

Construct a line graph of the results on the data sheet by plotting temperature on the horizontal axis and the average operculum movements on the vertical axis. Average the two readings for 22-26 °C and plot as one point. Plot the temperature points in numerical order.

## Notes on Graph Construction

1. A graph is a picture of the data. Plan the graph to properly fit the page; allow for numbers, labels, and margins.
2. The factor under the experimenter's control (usually time) is placed on the horizontal axis. The measured factor is placed on the vertical axis.
3. A graph should be self-explanatory:
   a. Title must be present
   b. Completely label each axis with factor, units, scale, etc.
4. If several curves are one graph, label each curve.

Name _____ Section _____

## Part II: Small Animal Metabolism Apparatus (oxygen consumption in a mammal)

### Material per four-eight students

One complete metabolism chamber
One thermometer
One mouse
Barium hydroxide, $Ba(OH)_2$

One cooler or refrigerator
Soap suds
Spoon

*Note:* The barium hydroxide bottle must be kept tightly closed so the chemical does not become inactive by absorbing $CO_2$ from the air.

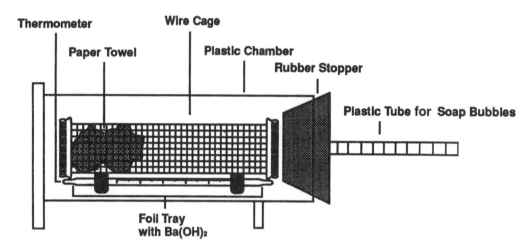

**Figure 29.1** Small Animal Metabolism Cage

### Method

1. Out of aluminum foil, fashion a tray to fit the bottom of the metabolism chamber. See Figure 29.1.
2. Put a crumbled paper towel in the back ¼ of the wire cage to restrict the movement of the animal.
3. Wear gloves when handling the animals; put the mouse in the wire cage.
4. Obtain a squeeze bottle of soap suds.
5. Wet the inside of the calibrated or marked tube with water.
6. Fill the aluminum foil tray with barium hydroxide and put it in the bottom of the plastic metabolism chamber. The $Ba(OH)_2$ will absorb all the $CO_2$ expelled by the animal and will allow for measurement of $O_2$ consumption only. Immediately close the barium hydroxide container so it does not become saturated with carbon dioxide from the air.
7. Attach the thermometer to the wire cage, and place it in the plastic chamber so the temperature is visible.
8. Insert the rubber stopper into the chamber.
9. The chamber is sealed by applying soap suds at the end of the tube.
   Note: It may take several minutes for the system to reach equilibrium and the soap bubble to begin moving down the tube. The soap bubble should be about 2 centimeters long.
10. Measure the amount of time in seconds required for the soap bubble to move the complete distance of the tube between the 5-0 mL marks on the tube. This distance is exactly equal to 5 mL. If a smaller volume tube is used, make the mathematical adjustments in the data table.

Name _____    Section _____

11. When the soap bubble has reached the inner end of the tube, immediately add another soap bubble to the outer end and begin a second trial. Two consistent trials should be obtained.
12. If inconsistencies persist after two trials, look for the following sources of error:
    a. leaks in the system
    b. insufficient or $CO_2$-saturated barium hydroxide
    c. dirty or blocked pipette
    d. failure to sufficiently wet the interior of the tube.
13. The experiment should be conducted at two different environmental temperatures, 5-15°C in the walk-in cooler, and 22-26°C room temperature.
14. With each change in temperature, the barium hydroxide must be replaced. Wash the used barium hydroxide down the drain with a large amount of water.
15. Conduct the first experiment at room temperature. Then remove the rubber stopper from the chamber. Remove the $Ba(OH)_2$. Place the chamber, the rubber stopper, and animal in the second environmental temperature, and allow the animal to stabilize for 15 minutes.
16. When working in the walk-in cooler, all experiments need to begin AT THE SAME TIME. DO NOT OPEN the cooler door as it will change the pressure in the test chamber. Post a guard outside the door to prevent other students from opening the door during the experiment.
17. Just before beginning the second experiment, add fresh barium hydroxide to the tray.
18. At the completion of the experiment, return the animal to its cage.
19. Wash the $Ba(OH)_2$ down the drain. Thoroughly wash and dry the chamber.

Name _____  Section _____

### TABLE 29.2
### Mouse Metabolism

| Temperature | Time (seconds) | Temperature | Time (seconds) |
|---|---|---|---|
| 5–15 °C<br><br>Exact Temp _____ | | 22–26 °C<br><br>Exact Temp _____ | |
| Total time in minutes | | Total time in minutes | |
| Total mL of $O_2$ consumed | | Total mL of $O_2$ consumed | |
| mL of $O_2$ consumed per minute | | mL of $O_2$ consumed per minute | |

1. Calculate the volume of $O_2$ consumed per minute by dividing the total number of mL consumed during two consistent runs by the length of time in minutes required for the two runs. If more than two runs have been conducted, select the two most consistent runs.
2. Construct a line graph of mouse metabolism by plotting temperature in numerical order on the horizontal axis and mL of $O_2$ consumed/minute on the vertical axis.

Name _____ Section _____

## Part III: Discussion

1. What is the relationship between increasing temperature and the operculum (breathing) rate of the fish?

2. Is a goldfish an ectotherm or an endotherm? What data support your answer?

3. Is a mouse an ectotherm or an endotherm? What data support your answer?

4. What similarities or differences did you observe in how a goldfish and a mouse responded to a cold environment?

5. Which of these animals would be better suited to live in an environment where the temperature fluctuated drastically? Cite a reason for your answer.

6. If you were too hot or too cold, what are some ways your body could regulate body temperature?

Name _____ Section _____

7. How does a goldfish deal with the problem of regulating body temperature?

8. What is the relationship between the temperature and the breathing rate in the mouse?

9. What is the relationship between the temperature and the breathing rate in the goldfish?

10. Compare the graph of the goldfish operculum movements to the graph of oxygen consumption of the mouse. Notice the slopes of the lines. Explain the meaning of these graphs.

Name _____   Section _____

# LAB 30

# Early Embryological Development

**Problem**

What are the early embryological stages of development?

**Objectives**

After completing this laboratory exercise, the student will be able to:

1. Identify the following stages of development:
   a. fertilization
   b. cleavage
   c. blastula
   d. gastrula.
2. Note the trend in cell number and cell size as development proceeds from fertilized egg to blastula.
3. Name the three germ layers of the gastrula.
4. Identify the major portions of the body which develop from each germ layer.

**Preliminary Information**

The general pattern of reproduction in most higher types of animals is the same. Sex cells (eggs and sperm) are produced by meiosis. Sperm and egg join in a process of **fertilization** forming a zygote. In some organisms, fertilization is external, usually occurring in water. In other organisms, fertilization occurs in the reproductive tract of the female. Following fertilization, the **zygote** begins a series of mitotic cell divisions. These early embryological divisions are called **cleavage** since the large egg is successively divided or cleaved into smaller cells. For a while, the cells divide so quickly that they do not have time to grow between divisions. Consequently, each new generation of cells becomes progressively smaller. Finally the speed of mitosis decreases allowing the subsequent generations of cells time to grow between divisions.

**Part I: Instructions**

1. In this lab exercise, the early stages of development will be examined. Obtain a prepared microscope slide of Starfish development, w.m. from the supply area. The "w.m." means whole mount. In a whole mount preparation, the entire specimen is placed on the slide rather than a slice or section of a specimen.
2. Scan the slide to locate the following stages of development.
3. Draw each stage of development in the space provided.

Name _____  Section _____

## Part II: Early Stages of Development

1. **Unfertilized egg**
   The unfertilized egg is a large single cell with a jelly-like coat around it. There is a distinct nucleus and a single nucleolus inside the nucleus.
2. **Zygote**
   In the fertilized egg (zygote), the nucleolus disappears. A fertilization membrane forms between the jelly coat and the surface of the egg.
3. **Two-cell stage**
   The single-celled fertilized egg pinches into two cells. Each new cell is half the size of the egg cell.
4. **Four-cell stage**
   Each cell of the two-cell stage pinches into two, forming a four-cell structure. Each new cell is half the size of its parent cell.
5. **Eight-cell stage**
   Each cell of the four-cell stage simultaneously divides.
6. **Sixteen-cell stage**
   Each cell of the eight-cell stage simultaneously divides.
7. **Morula or thirty-two-cell stage**
   The morula is a solid ball of 32 cells looking like a raspberry.
8. **Blastula**
   The blastula is a hollow ball of cells with a central cavity called the blastocoel.
9. **Gastrula**
   The gastrula stage begins when the cells at one spot on the hollow ball stage begin to migrate into the central cavity. This appears as a depression on the surface of the blastula. Imagine pushing your finger into a tennis ball. This is the way the gastrula will appear. The infolding of the outer surface will continue until the blastocoel is eliminated. At this point, two of the three embryonic germ layers can be seen, the ectoderm and the endoderm. The different cell layers of the gastrula are called germ layers because they contain the initial cells which will later develop into specific body structures. Soon the third germ layer, the mesoderm, will form between the first two layers.
   The three embryonic germ layers are:
   a. the ectoderm or outer layer
   b. the endoderm or inner layer which has just infolded
   c. the mesoderm or middle layer which forms between the ectoderm and endoderm.
   The ectoderm will form the skin and nervous system. The endoderm will form primarily the digestive tract. The mesoderm will form primarily muscle and bone.
10. **Bipinnaria larva**
    This larva stage is becoming more complex and beginning to show lateral symmetry. The bulging oral lobe and the mouth are generally visible toward one end with a large round stomach toward the other end. The stomach will have a tube attached to it that ends at the anus.

**TABLE 30.1**
Early Embryological Development

| Unfertilized egg | Zygote |
|---|---|
| | |

| Two-cell stage | Four-cell stage |
|---|---|
| | |

| Eight-cell stage | Sixteen-cell stage |
|---|---|
| | |

| Morula stage | Blastula stage |
|---|---|
| | |

| Gastrula stage | Bipinnaria larva |
|---|---|
| | |

Name _____  Section _____

**Part III: Discussion**

1. How can you distinguish an unfertilized egg from a fertilized egg?

2. With each cleavage, what happens to the number of cells?

3. What is the trend in cell size as development proceeds from zygote to blastula?

4. Construct a table listing the three germ layers of the gastrula and the major body portions formed from each layer.

Name _____    Section _____

# LAB 31

# Effect of Hormone on Plant Growth

*Note:* The seeds must be soaked for 60-80 minutes before planting.

## Problem

What effect will gibberellins have on plant growth in normal and dwarf pea plants?

## Objectives

After completing this exercise, the student will be able to:

1. Identify the role of a hormone in plants.
2. Discuss the effect gibberellic acid has on the growth of normal and dwarf pea plant stems.
3. Construct a table and a graph from the data obtained.
4. Explain the use of a control in an experiment.
5. Compare the germination rates between normal and dwarf pea plants.

## Preliminary Information

Maintenance and growth in plants is under the regulation of plant hormones. Like animal hormones, plant hormones are synthesized in one part of the organism but have an effect in a different part of the organism. Hormones are effective in minute quantities. Gibberellin or gibberellic acid is one type of plant hormone.

Gibberellins specifically promote cell enlargement but do not cause curvatures of the stem. The most common effect attributed to application of gibberellins is the extreme elongation of the stem, but it is also involved in flowering response and breaking dormancy of seeds.

## Hypothesis

In this experiment gibberellins will be applied to both dwarf and normal pea plants. Before beginning the experiment it is best to give some thought to how you expect the experiment to turn out. Use the preliminary information and develop a probable outcome or hypothesis. If in the end, the hypothesis is found to be incorrect, this in no way decreases the validity of the hypothesis.

Hypothesis: _____

_____

Name _____   Section _____

## Part I: Stem Growth

### Materials and Methods

The following will be needed for each team of four or as the instructor directs:

    20 presoaked normal pea seeds      *Note* Seeds should be presoaked for
    20 presoaked dwarf pea seeds     60-80 minutes.
    1 planting container with soil

Divide the tray into fourths with strips of masking tape and label as follows:

    Normal control         Normal treated
    Dwarf control          Dwarf treated

Record all names and the section number on tape and apply to the tray.
Plastic wrap
Enough soil for one tray

### Procedure

1. Label the tray as suggested under materials section and add soil.
2. Plant ten normal pea seeds in each section labeled "normal control" and "normal treated." Plant 10 dwarf pea seeds in each section labeled "dwarf control" and "dwarf treated."
3. Poke holes one cm deep with a pencil. Place seeds in the holes Cover seeds with soil.
4. Add enough water to moisten all of the soil.
5. Check daily that plants are moist enough. The soil should feel damp to the touch but not wet. Overwatering or underwatering can effect how many seeds will germinate.
6. Particularly make sure all plants are watered on Friday afternoon and Monday morning.
7. Cover trays loosely with plastic wrap: Fold the plastic wrap back at alternate corners to leave a 2-3 cm opening for air circulation. Remove the plastic when plants are 1-2 cm high.
8. Place trays under plant growth lights.
9. Seven days after planting:
   a. Number each plant for identification with a small durable paper collar or flags made of tape and toothpicks.
   b. Measure the height in millimeters from the surface of the soil to the tip of the shoot of each plant.
   c. Record all initial measurements on data sheets under Day 0.
10. Spray the treated normal and treated dwarf peas with gibberellic acid. Spray the leaves and apex until droplets form. Avoid overspraying. **Caution**. Cover control peas so they are protected from the spray.
11. Measure all plants and spray treated plants during each class for two weeks. Record all data.
12. Calculate and record the percent of increase for each group for each day (formula is with data sheet).
13. Graph the effect of gibberellic acid on stem growth by plotting time in days on the horizontal axis and percent of increase in length on the vertical axis.

Name _____     Section _____

## Part II: Data

**TABLE 31.1**
Effects of Gibberellic Acid on Stem Growth

| Type | Plant Number | Initial Length (mm) Day 0 | Length in mm | | | | |
|---|---|---|---|---|---|---|---|
| | | | Day ___ | Day ___ | Day ___ | Day ___ | Day ___ |
| Normal control | 1 | | | | | | |
| | 2 | | | | | | |
| | 3 | | | | | | |
| | 4 | | | | | | |
| | 5 | | | | | | |
| | 6 | | | | | | |
| | 7 | | | | | | |
| | 8 | | | | | | |
| | 9 | | | | | | |
| | 10 | | | | | | |
| | average | | | | | | |
| | % increase | ///// | | | | | |
| Dwarf control | 1 | | | | | | |
| | 2 | | | | | | |
| | 3 | | | | | | |
| | 4 | | | | | | |
| | 5 | | | | | | |
| | 6 | | | | | | |
| | 7 | | | | | | |
| | 8 | | | | | | |
| | 9 | | | | | | |
| | 10 | | | | | | |
| | average | | | | | | |
| | % increase | ///// | | | | | |

Note: Always record the number of days that have passed *since* the initial length measurement was taken.

% Increase for Day 2 = $\dfrac{\text{Average length Day 2} - \text{Average initial length}}{\text{Average initial length}}$ X 100

Use Average initial length for all calculations, not the average from the previous day.

Repeat % Increase calculations for each day data was recorded.

Name _____  Section _____

TABLE 31.1 (continued)

| Type | Plant Number | Initial Length (mm) Day 0 | Length in mm | | | | |
|---|---|---|---|---|---|---|---|
| | | | Day ___ | Day ___ | Day ___ | Day ___ | Day ___ |
| Normal treated | 1 | | | | | | |
| | 2 | | | | | | |
| | 3 | | | | | | |
| | 4 | | | | | | |
| | 5 | | | | | | |
| | 6 | | | | | | |
| | 7 | | | | | | |
| | 8 | | | | | | |
| | 9 | | | | | | |
| | 10 | | | | | | |
| | average | | | | | | |
| | % increase | ///// | | | | | |
| Dwarf treated | 1 | | | | | | |
| | 2 | | | | | | |
| | 3 | | | | | | |
| | 4 | | | | | | |
| | 5 | | | | | | |
| | 6 | | | | | | |
| | 7 | | | | | | |
| | 8 | | | | | | |
| | 9 | | | | | | |
| | 10 | | | | | | |
| | average | | | | | | |
| | % increase | ///// | | | | | |

## Germination Rate

Calculate percentage of germination for normal and dwarf plants.

$$\frac{\text{Number of seeds germinated}}{\text{Number of seeds planted}} \times 100 = \text{Percent of germination}$$

Percent of germination for normal plants = _____

Percent of germination for dwarf plants = _____

Name _____  Section _____

## Part III: Graph Effect of Gibberellic Acid

1. Before beginning graph construction, calculate the percent of increase of length for each day. The graph should be constructed on a sheet of graph paper. Remember to include a **title** and **label** each axis.
2. Plot all curves on the same graph and label each curve.
3. Plot time on the horizontal axis and percent of increase of length on the vertical axis. In plotting time, remember to include days over the weekend even though no measurements were taken on these days.

## Part IV: Discussion

1. Why bother with the control plants if they are not going to be sprayed with gibberellic acid?

2. Do the normal pea plants respond to the treatment of gibberellins in the same way as the dwarf plants? Why or why not?

3. Did you observe any difference in germination rates between normal and dwarf plants? What is the significance of this?

Name _____  Section _____

4. Has the experiment given you any idea about the role of gibberellins in a normal plant?

5. From your data, what can you conclude as to the probable causes of dwarf plants?

Name _____  Section _____

# LAB 32

# Chlorophyll Extraction and Separation

**Problem**

Can the presence and function of plant pigments be determined?

**Objectives**

After completing the exercise, the student will be able to:

1. Discuss the method used to extract plant pigments.
2. Identify the number of pigments found in a leaf.
3. Explain the process of paper chromatography.
4. Relate the color of plant pigments in a leaf to the process of photosynthesis.
5. Identify a simple test to detect the presence of starch.

**Preliminary Information**

The most predominant plant pigment is chlorophyll. Chlorophyll is necessary for photosynthesis. It gives the leaf its green color because certain wavelengths of light are absorbed and green wavelengths are reflected by the plant. Our eyes see this reflected green color.

Chlorophyll often masks the presence of other plant pigments. In the autumn, the chlorophyll is no longer produced and other plant pigments can be seen. This is why the marvelous array of reds and yellows appear in the fall. These pigments can be seen in plants with variegated leaves. Variegated leaves have patches of white, yellow or pink which lack the green pigment, chlorophyll.

Photosynthesis is the biochemical process that results in the production of a simple sugar. Some of the simple sugar produced in the leaf is converted into the more complex form of starch. The presence of starch can be tested by adding iodine. The starch will appear brown or purple in the leaf.

Paper chromatography is a technique that can be used to separate plant pigments. This process involves dissolving the pigments in a solvent then allowing the pigments to move up a strip of chromatography paper with the aid of a second solvent. Since the pigments are of different molecular weights, they move up the paper at different speeds and can thus be separated from each other.

Name _____   Section _____

## Part I: Pigment Extraction and Separation

### Materials

Each group should obtain the following materials from the supply area:

One strip of chromatography paper. Handle by the edges only.
One large test tube.
One cork or rubber stopper to fit test tube.
One wire hook (paper clip) *Note:* This will be used to suspend the chromatography paper inside the test tube. See Figure 32.1.
Fine tipped pipette.
Mortar and pestle.
One spinach leaf.
Approximately 6 mL of acetone in the mortar.

*Note*: One dropper full of acetone is about 1 mL.

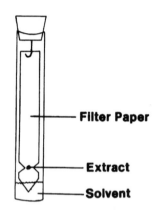

**Figure 32.1** Chromatography Apparatus

### Procedure

**Note: Wear safety glasses.**

Note: Refer to Figure 32.1 for the assembled chromatography apparatus.

1. Put the wire hook through the stopper.
2. Do not touch the face of the chromatography paper.
   Handle it by the edges.
   Trim the chromatography paper to fit into the test tube. When the paper is suspended, it should hang straight without touching the sides or bottom of the test tube.
   Trim the bottom of the chromatography paper to a V-shaped point. Approximately 2 cm from the bottom, cut a V-shaped notch out of each side of the chromatography strip.
3. Prepare a chlorophyll extract by vigorously grinding spinach leaves in approximately 6 mL of acetone using the mortar and pestle. Note: A full squeeze from a dropper bottle equals about 1 mL.
4. Using a fine tipped pipette, apply a VERY THIN BAND or a DROP of extract between or slightly below the notches in the chromatography paper.
5. Allow the extract to dry completely. Blowing the paper will help it dry faster.
6. Repeat steps 4 and 5 until 10 coats of extract have been applied.
7. Add 3-5 mL of petroleum ether-acetone solvent to the test tube.

Name _____ Section _____

8. Attach the strip of chromatography paper to the stopper and place it in a test tube. **Caution**. The band of extract must not be submerged in the solvent. Place the tip of the paper 1 cm into the solvent. The paper must hang straight and not touch the sides of the test tube.
9. Carefully place the test tube in a large beaker containing some crumpled paper towels to hold the test tube straight.
10. Observe at about 3 minute intervals until 4 color bands are visible. When a thin bright yellow band reaches the top of the strip, remove the strip.
11. Dispose of the petroleum ether-acetone solvent in the marked waste container.
12. On the data sheet, diagram and label the pigments on the chromatography paper when the separation of pigments is complete, or staple the strip to the data sheet.

## Part II: Pigments Responsible for Photosynthesis

### Materials

Each group should obtain the following materials from the supply area:

One leaf
500 mL or larger beaker
Large test tube
Alcohol, fill test tube ½ full
Petri dish
Iodine in dropper bottle
Hot plate

### Procedure

**Note: Wear safety glasses.**

1. Using a hot plate, begin heating half a large beaker of water to boiling.
2. Obtain a large test tube and fill it ½ full of alcohol.
3. Add the leaf to the alcohol.
4. Place alcohol test tube with leaf in the beaker of lightly boiling water.
5. Heat alcohol test tube for 15 minutes. *Note:* You may need to add more alcohol to the test tube if too much alcohol boils away.
6. Fill the petri dish ⅓ full of water.
7. Remove the leaf, place it on several layers of paper towel and add 10-15 drops of iodine solution directly to the leaf.
8. Transfer the leaf to the petri dish with water.
9. Sketch the leaf indicating the stained areas.

   **Note:** Iodine stains starch a dark color, so the darkly stained portions of the leaf indicate the presence of starch.

10. Wash and return glassware. Clean lab table.

Name _____    Section _____

**Part III: Data and Discussion**

1. Paper Chromatography: Either attach the chromatography strip or draw the chromatography strip in the space below. Pigments that could be present are:

   Chlorophyll a—blue-green color
   Chlorophyll b—yellow-green color (least soluble)
   Xanthophyll—pale yellow color
   Carotene—bright-yellow color (most soluble)

   Label the pigments present on your strip or your drawing.




2. What is the most predominant pigment present in the spinach leaf? Give evidence to support your answer.




3. What colors were the pigments extracted from the spinach leaf? Identify the pigments.




4. Carotene is the most soluble pigment and chlorophyll b is the least soluble. Where should each of these be located? Do the data support these?

Name _____  Section _____

5. Pigments Responsible for Photosynthesis

   Diagram the iodine-stained leaf
   in the space below

6. Why was the leaf placed in the iodine?

7. Iodine stains starch a dark purple color. Did the leaf test evenly for the presence of starch? Why or why not?

8. Since photosynthesis produces sugar which can then be stored as starch, what pigment can be associated with photosynthesis? Give evidence to support your answer.

Name _____  Section _____

# LAB 33

# Diversity of Plants and Fungi

**Problem**

What characteristics are used to identify different plants and fungi?

**Objectives**

After completing this exercise, the student will be able to:

1. List the classification from kingdom through species in proper sequence.
2. Recognize the characteristics used to differentiate among the various groups of plants and fungi.
3. List the characteristics of each plant or fungus group.
4. Identify an unknown organism into the proper kingdom, phylum, or class.
5. List characteristics and be able to identify examples of the following groups:

    Kingdom Fungi
    Kingdom Monera
    Kingdom Protista
      Division Chlorophyta
    Kingdom Plantae
      Division Bryophyta
      Division Pteridophyta
      Division Gymnospermae or Pinophyta
      Division Angiospermae or Magnoliophyta
        Class Monocotyledonae
        Class Dicotyledonae

**Preliminary Information**

Over a million species of animals and over 350,000 species of plants and fungi are now known. Most of the early biologists were naturalists and they spent much of their time identifying and describing animals and plants. But it was not enough to describe these organisms. A system was needed to organize plants and animals in a logical and meaningful way.

Many different ways to classify organisms have been used. For example, plants have been classified according to their size; placing all tall plants (like trees) in one group, all intermediate sized plants (like shrubs) in another group, and short plants (like most houseplants) in a third group. Or they have been classified according to habitat; separating field plants, marsh plants, desert plants, forest plants, ocean plants, etc., from each other. Anatomical relationships among plants have been used. Using the reproductive structure of plants is the most accurate and consistent way to classify plants.

Name _____  Section _____

## Part I: Natural System of Plant Classification

The classification system of plants that is most widely accepted today is based on a *natural system* in which the groupings of plants show how closely related they are. This system is broken down as follows:

    Kingdom                          Country
      Division                      State
        Class                        Town
          Order                   Street number
            Family               House number
               Genus (plural, genera)   Last name
                  Species             First name

Each category in this hierarchy is a collective unit containing one or more groups from the next lower level in the hierarchy. Thus a phylum is a group of related classes; a class is a group of related orders; an order is a group of related families, etc.

## Part II: How to Use a Key

Most keys begin with some obvious characteristic and present you with at least two choices. In the kingdom key you must first decide if the organism is single or multicellular. A single cell is so small that you need a microscope to see it whereas multicellular organisms can be seen without a microscope. Do not confuse a colonial organism made of clusters or a chain of similar cells with a multicellular organism. If it is single-celled you are referred to item number 2 for the next two choices. Select the item that best fits the organism in question. However, if the organism is multicellular you are referred to item number 3 where you must decide if it produces its own food or not.

### Key to Kingdoms

1a. Single-celled organisms or colonies of similar cells; usually microscopic......................2
1b. Multicellular organisms; body composed of tissues or layer of cells; usually macroscopic..........3

2a. Cells with a nucleus ................................................................... **Kingdom Protista**
2b. Cells without nucleus .................................................................. **Kingdom Monera**

3a. Produces own food, usually green; lacks self-mobility........................**Kingdom Plantae**
3b. Does not produce own food; generally some portion of life cycle mobile ....... **Kingdom Animalia**
3c. Does not produce own food; never mobile; often mushroom-shaped............... **Kingdom Fungi**

## Part III: Recognition of Plant and Fungus Characteristics

In the following exercises you will become familiar with some major groups of plants and fungi. You will study the following groups:

1. Fungi
2. Green algae
3. Mosses
4. Ferns

Name _____    Section _____

5. Seed plants
   a. Evergreens
   b. Flowering plants
      (1) Monocots
      (2) Dicots

Notice that you will be looking at the subgroups of seed plants in detail. Most of the plants around us belong to this group.

The placement of a plant into one of these groups depends upon the presence or absence of certain characteristics. Before you begin to classify plants and fungi yourself, you must be able to recognize these characteristics.

## *Examination of Plant and Fungus Characteristics*

On the demonstration table you will find materials which will help you to become familiar with certain plant terms.

### A. Chlorophyll

The presence of the green pigment chlorophyll gives a green color to the organism. Look at a living plant and a fungus on the demonstration table.

What color is the plant? _____

Does it have chlorophyll? _____

What color is the fungus? _____

Does it have chlorophyll? _____

### B. True Leaves and Stems

Vascular bundles or veins are groups of specialized conductive tubes which move water, food, minerals and hormones through a plant. **The presence of vascular tissue is the criterion for determining if a plant has true leaves or stems.** The vascular bundles of a leaf are the veins of the leaf. A single midrib by itself is not evidence of vascular bundles. Observe the leaflike structures of the material on the demonstration table. To establish if true leaves are present, hold the leaf up to the light. If you observe veins running throughout the leaf, it is a true leaf. Or observe the cut end of the stem of a celery stalk and a mushroom. If round dots are present, these are vascular bundles or veins. **If a stem is woody, it has a concentration of vascular tissue** which appear as rings in cross-section.

Which organisms at this demonstration area possessed vascular bundles? _____

_____

Which organisms at this demonstration area did not possess vascular bundles? _____

_____

Name _____   Section _____

Examine prepared slides of a cross section through a stem of corn *(Zea)* and mushroom *(Coprinus)* which are available at the demonstration table. Look for the presence of vascular bundles that look like monkey faces. Refer to the Figure 33.1 of cross section of corn stem and locate the vascular bundles.

Did you locate vascular bundles in corn? _____

Did you locate vascular bundles in mushrooms? _____

C. **Seeds and Spores**

Most plants reproduce either by spores or seeds. *Spores* are unicellular; *seeds* are multicellular and are composed of a seed coat, an embryo, and stored food. Spores are produced by fungi, green algae, liverworts, mosses, and ferns. Seeds are produced by the evergreens and flowering plants. At the demonstration table observe the materials labeled seeds and spores.

What is the size difference between spores and seeds? _____

_____

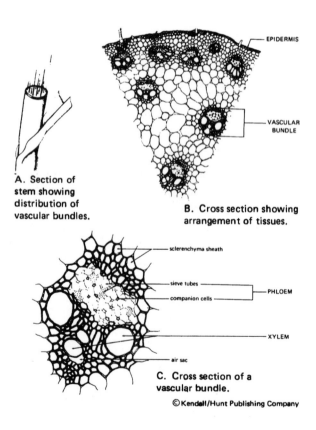

Figure 33.1 Structure of a Monocot (Corn) Stem

Name _____  Section _____

D. **Cotyledons**

All seeds either separate into two halves (dicotyledons) or do not separate easily (monocotyledons). Obtain a soaked bean and corn seed from the supply area. Examine each by taking the seeds apart.

Which is the monocotyledon seed? _____

Which is the dicotyledon seed? _____

Plants producing monocotyledon seeds are called *monocots* and plants producing dicotyledon seeds are called *dicots*.

E. **Flower Parts**

All covered seeds are borne in modified structures of flowers. Observe a typical flower (Figure 33.2) and notice the circle of brightly colored *petals*, the pollen producing *stamens*, and the vase-shaped *pistil*. Stamens are the male reproductive parts and the pistil is the female reproductive part of the flower. A pollen grain from the same flower or a different flower of the same species will fertilize the egg cell within the base of the pistil. The fertilized egg will develop into an embryo. The embryo and its large mass of stored food is called a seed. Thus the seed is completely enveloped by the pistil. It is the pistil of the flower which eventually develops into *fruit*, usually greatly enlarging in the process.

How many petals are found in the tulip flower? _____

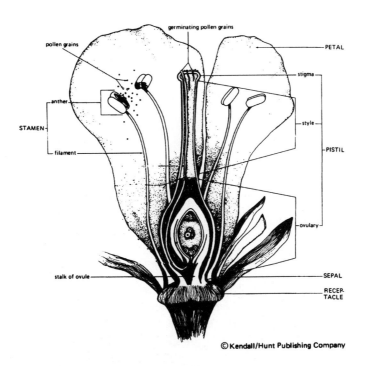

**Figure 33.2** Generalized Structure of a Dicot Flower

Name _____   Section _____

How many petals are found in the dicot flower? _____

The flower parts (petals, stamens, pistils) of monocots occur in three, or multiples of three. The flower parts of dicots occur in four or five, or multiples of four or five.

F. **Leaf Venation**

Leaves having veins which lie parallel to one another from the base to the tip of the leaf are *parallel-veined* leaves. Leaves having veins which split off from the main midrib vein so that the venation resembles a net are *net-veined* leaves. Examine the leaves of tulip and violet or substitute plants.

Does the monocot have parallel-veined leaves or net-veined leaves? _____

Does a dicot have parallel-veined leaves or net-veined leaves? _____

**Parallel-veined leaves indicate a monocot and net-veined leaves indicate a dicot.**

G. **Summary of Plant Characteristics**

A green color indicates the presence of _____.

True leaves are indicated by the presence of _____.

Wood is _____.

If a seed, after removing the seed coat, splits into two halves while you are examining it, the plant that produced this seed could be called a _____.

A _____ matures and ripens into a fruit containing seeds.

If a flower has three petals, the plant could be called a _____.

If a plant has net venation and five petals, it could be called a _____.

**Part IV: Classification of Unknown Plants**

Using the taxonomic key at the end of the exercise, classify the given plant or parts of a plant from your plant set into the appropriate categories.

**Use of Keys for Plant Kingdoms, Division and Classes**

The type of key that will be utilized is referred to as a dichotomous or branching key, which means at every level of the key you always have at least two choices. Read the descriptions and decide which one best fits the specimen you are keying. To the right of each description you are referred by number to the next part of the key or the classification name of the kingdom, division, or class is given.

Name _____    Section _____

Begin by locating the 11 specimens at the demonstration table. Classify each specimen and then check with your instructor to make sure you have classified everything correctly. If any are wrong, go through the key again to locate your error.

This will be valuable practice for the practical quiz that you will have to take. The practical quiz will consist of 10 unknown specimens that you will have to identify to kingdom, division, and class *without* the use of the key.

## Taxonomic Key for Plants, Fungi and Protists

1a. Chlorophyll absent; no true roots, stems, and leaves; often mushroom-shaped . . . . . **Kingdom Fungi**
1b. Chlorophyll present . . . . . . . . . . . . . . . . . . . . . . . . . . . . . . . . . . . . . . . . . . . . . . . . . . . . . . . go to 2

2a. Single-celled organisms or colonies of single cells: aquatic;
usually microscopic . . . . . . . . . . . . . . . . . . . . . . . . . . . . . . **Kingdom Protista, Division Chlorophyta**
2b. Multicellular organisms . . . . . . . . . . . . . . . . . . . . . . . . . . . . . . . . . . . . . **Kingdom Plantae:** go to 3

3a. Vascular tissue absent . . . . . . . . . . . . . . . . . . . . . . . . . . . . . . . . . . . . . . . . **Division Bryophyta**
3b. Vascular tissue present . . . . . . . . . . . . . . . . . . . . . . . . . . . . . . . . . . . . . . . . . . . . . . . . . . go to 4

4a. Leaves with large number of lobes indented to the midrib;
Spores on under surface of leaves . . . . . . . . . . . . . . . . . . . . . . . . . . . . . . . . . . **Division Pteridophyta**
4b. Needlelike leaves; seeds produced in cones . . . . . . . . . . . . . . . . . . . . . . . **Division Gymnospermae**
4c. Leaves broad or narrow; seeds produced in flowers . . . . . . . . . . . . . **Division Angiospermae:** go to 5

5a. Parallel-veined leaves; flower parts in threes; seeds with one cotyledon . . . . **Class Monocotyledonae**
5b. Net-veined leaves; flower parts in four or fives; seeds with two cotyledons . . . . . **Class Dicotyledonae**

**TABLE 33.1**
Plant Classification Using a Taxonomic Key

Note: Each specimen may not have all the categories.

| Specimen Number | Common Plant Name | Kingdom | Division | Class |
|---|---|---|---|---|
| 1. | Fern | | | |
| 2. | Spider Plant | | | |
| 3. | Pine branch w/cone | | | |
| 4. | Oak leaf and acorn | | | |
| 5. | Moss | | | |
| 6. | Green algae | | | |
| 7. | Corn leaf w/corn kernels | | | |
| 8. | Mushroom | | | |
| 9. | Grass | | | |
| 10. | Tulip flower | | | |
| 11. | Violet flower | | | |

Name _____  Section _____

# LAB 34

# Population Ecology

**Developed by Karen Borgstrom**

**Problem**

How do survivorship and mortality rates impact population ecology?

**Objectives**

After completing this lab exercise, the student will be able to:

1. Construct life tables using basic population ecology equations and preliminary data on a hypothetical population.
2. Construct survivorship curves given population data from a life table.
3. Distinguish between the three basic types of survivorship curves observed among natural populations.
4. Simulate populations and gather data to be tabulated in a life table and plotted on a survivorship curve.
5. Compare survivorship curves of a developed and underdeveloped nation.

**Preliminary Information**

Populations can be characterized in a number of ways. One way an ecologist can characterize a population is based on mortality rates, or, conversely, survivorship. Since all members of a population are not equal, the life span of each member can and does vary tremendously: some will die shortly after birth and others will survive to be the eldest of a group.

An ecologist can statistically represent the mortality and survivorship of a given population using a life table. Health and life insurance companies have long used life tables to determine rates for members of a particular cohort. A **cohort** consists of a group of individuals in a population who were born at the same time and are researched until the last member of the cohort dies. Among scientists, life tables are used in public health, conservation of endangered species, forestry, management of pests, and ecology. Even governments have an interest in knowing the number of individuals at an age for military service, education or who might be drawing on Social Security in 2030.

**Part I: Construction of a Life Table**

Although rather simple to construct, the basic formula used in determining survivorship can be confusing. When using a life table one begins with all members of a cohort just beginning life (i.e., time interval or age is equal to zero). An 'x' is used to represent a time interval or age, therefore, initially $x = 0$. The number of deaths during an age interval is represented as '$D(x)$' and the number of survivors at the beginning of an age interval is represented as '$S(x)$.' Thus, in Life Table 34.1, $S(x=0)$ is 200 (i.e., the original number of individuals in the cohort) and $D(x=0)$ is 20 (i.e., the number of individuals who died during the first age interval). Therefore, to calculate S(x=1) (i.e. the number of individuals alive at the beginning of the second age interval) use the following formula:

$$S(x + 1) = S(x) - D(x)$$

Thus, we can determine S(x=1) by S(0) – D(0) or S(1) = 200 – 20 = 180. As you can see all data can be filled in given the deaths during each age interval and using the above equation. Note, the cohort was researched until all members of the population perished.

**TABLE 34.1.**
Life Table for a Hypothetical Population

| x | S(x) | D(x) |
|---|------|------|
| 0–.99 | 200 | 20 |
| 1–1.99 | 180 | 40 |
| 2–2.99 | 140 | 60 |
| 3–3.99 | 80 | 80 |
| 4–4.99 | 0 | 0 |

A life table for a hypothetical population. The members of this population had a life span of five age intervals. Note, x=0 represents the time from birth to age interval one and D(x=0) represents the number of deaths during that time. In this population there were 200 members originally and the last member died between the age interval of 3-4.

Another factor used in evaluating a cohort is the mortality rate. Since mortality implies data based on members who are deceased, and we are referring to survivors, we will call this rate the **age specific survivorship or $\ell(x)$**. Age specific survivorship is the *percentage* of individuals from the original cohort who have survived to a particular age interval. Age specific survivorship can be calculated using the simple formula:

$$\ell(x) = S(x) / S(0)$$

For age 0, $\ell(x)$ is always equal to 1.00 or 100%. We can calculate age specific survivorship for the data from Table 34.1.

**TABLE 34.2.**
Hypothetical Population Age Specific Survivorship

| x | S(x) | D(x) | $\ell(x)$ |
|---|------|------|-----------|
| 0–.99 | 200 | 20 | 1.00 |
| 1–1.99 | 180 | 40 | 0.9 |
| 2–2.99 | 140 | 60 | 0.7 |
| 3–3.99 | 80 | 80 | 0.4 |
| 4–4.99 | 0 | 0 | 0 |

Life table for a hypothetical population which includes age specific survivorship for data from Table 34.1.

Name _____  Section _____

Keeping in mind the two formulas used:

$$S(x + 1) = S(x) - D(x) \text{ and } \ell(x) = S(x) / S(0)$$

answer the following questions by completing the life table below.

**Table 34.3**
Hypothetical Population Age Specific Survivorship

| x | S(x) | D(x) | $\ell(x)$ |
|---|---|---|---|
| 0–.99 | 100 | 40 | 1.00 |
| 1–1.99 | 60 | 25 | 0.60 |
| 2–2.99 | 35 | 15 | |
| 3–3.99 | | 10 | |
| 4–4.99 | | 10 | |
| 5–5.99 | | | |

How many members were in the cohort originally? _____

What is the value of D(x=2)? _____

What is the value of S(x=3)? _____

What is the value of $\ell$(x=4)? _____

## Part II: Survivorship Curves

Tables are useful for illustrating numerical data, but may require study to illustrate trends. Graphical presentations better illustrate trends which may be advantageous when studying population ecology. In fact, life table data is often plotted to form a survivorship curve. A survivorship curve will plot age specific survivorship, $\ell(x)$, against a time interval or age of the cohort, x. In addition, a graph will allow one to compare survivorship curves of two or more populations regardless of the original population size.

Figure 34.1 below is a survivorship curve for the data from Table 34.2. Note, $\ell(x)$ is plotted on the y-axis and x is plotted on the x-axis.

Name _____  Section _____

## Survivorship Curve for Data in Table 34.2.

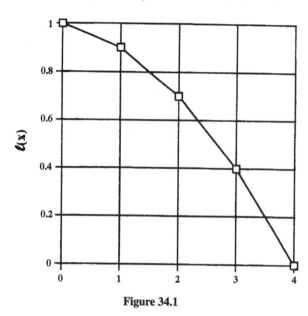

Figure 34.1

Construct a survivorship curve on the graph in Figure 34.2 using the data from Table 34.4.

**TABLE 34.4**
Life Table

| x | $\ell(x)$ |
|---|---|
| 0 | 1.0 |
| 1 | 0.3 |
| 2 | 0.2 |
| 3 | 0.1 |
| 4 | 0 |

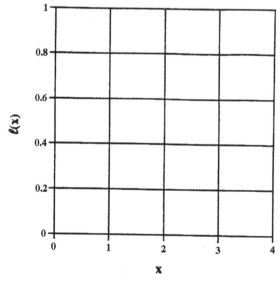

**Figure 34.2** Graph of Table 34.4.

Name _____                              Section _____

## Part III: The Three Basic Patterns of Survivorship Curves

Below is a description of each of the three basic types of survivorship curves. Keep in mind these curves are based on theoretical data while actual population data may model a certain type but may not exactly fit the curve.

Type I:   This type of curve is found in populations which are well adapted to their environment, have few offspring, and provide extensive parental care. In such populations most members live long lives and then mortality abruptly increases very late in life when most members perish. This type of curve is often seen in large mammals, including humans, and annual plants.

Type II:  This type of curve is reflected in populations which have a constant death rate at all ages. Although not very common in nature, this type of curve may be observed in populations of birds, lizards, and small mammals.

Type III: This type of curve is usually seen in populations which have a large number of offspring and little parental care. In such populations there is a high mortality rate early in life. Those members who survive the early years, however, live relatively long lives. Such a curve is seen in fish, invertebrates, and perennial plants.

The graph below shows typical survivorship curves for each of the above survivorship types discussed.

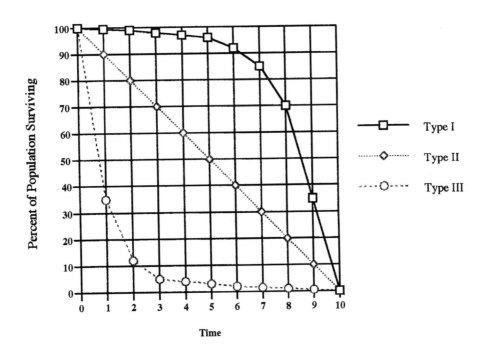

**Figure 34.3** Survivorship Curves

Name _____   Section _____

## Part IV: Population Simulations

A hypothetical population may be studied using some common everyday objects. Three population simulations will be demonstrated using soap bubbles and dice.

### Population Simulation #1

1. Obtain a bottle of soap bubbles.
2. One student will serve as the "bubble blower," one as "timer," and the remaining students will be "bubble counters." The "timer" should count each second out loud so the "bubble counters" know how long the bubble lasted. Each time, the "bubble blower" should blow about three bubbles **high into the air** for each "bubble counter."
3. A total of 50 bubbles should be made and their mortality data recorded in the table for Population Simulation #1. Each bubble represents a single member of the population of 50 soap bubbles. For example, if a bubble is produced and it "pops" at 6 seconds, make a tally mark at interval six in the table.
4. Continue this process until all 50 bubbles have been tallied.
5. Do the calculations necessary to complete the data table for Population Simulation #1.
6. Plot the data x versus $\ell(x)$ on graph paper.
   **Note**: Data from Population Simulations #1, #2, and #3 are all plotted on the same graph. Plot time on the horizontal axis and $\ell(x)$ on the vertical axis. Label each curve.

### Population Simulations #2

1. Obtain 50 dice and roll all the dice at once.
2. Each die represents a single member of the population of 50. For a score of 3, 4, 5 and 6 the individual survives time interval one. For a score of 1, the individual dies of heart disease and the die is removed. For a score of 2, the individual dies of AIDS and the die is removed.
3. Record the data in the table for Population Simulation #2.
4. Continue to roll the surviving dice and record results until there are less than three dice remaining.
5. Do the calculations necessary to complete the data table for Population Simulation #2.
6. Plot the data x versus $\ell(x)$ on graph paper.

### Population Simulation #3

1. Imagine a vaccine for AIDS has been discovered. Thus, repeat the procedure in #2, except for rolls of 2, 3, 4, 5 and 6 the individual survives and for a roll of 1 the individual will die of heart disease.
2. Again, continue to roll the surviving dice until there are less than three remaining.
3. Record data in the table for Population Simulation #3.
4. Do the calculations necessary to complete the data table for Population Simulation #3.
5. Plot the data x versus $\ell(x)$ on graph paper.

Name _____  Section _____

**TABLE 34.5**
Table for Population Simulation #1.

| x (age in sec) | Tally (number dying during interval) | S(x) | D(x) | $\ell(x)$ |
|---|---|---|---|---|
| Population Simulation #1—Soap Bubble Data ||||| 
| 0–.99 | | | | |
| 1–1.99 | | | | |
| 2–2.99 | | | | |
| 3–3.99 | | | | |
| 4–4.99 | | | | |
| 5–5.99 | | | | |
| 6–6.99 | | | | |
| 7–7.99 | | | | |
| 8–8.99 | | | | |
| 9–9.99 | | | | |
| 10–10.99 | | | | |
| 11–11.99 | | | | |
| 12–12.99 | | | | |
| 13–13.99 | | | | |
| 14–14.99 | | | | |
| 15–15.99 | | | | |
| 16–16.99 | | | | |
| 17–17.99 | | | | |
| 18–18.99 | | | | |
| 19–19.99 | | | | |
| 20–20.99 | | | | |

Name _____   Section _____

**TABLE 34.6**
Table for Population Simulation #2.

| x | Tally 1 | Tally 2 | S(x) | D(x) | $\ell(x)$ |
|---|---|---|---|---|---|
| 0–.99 | | | | | |
| 1–1.99 | | | | | |
| 2–2.99 | | | | | |
| 3–3.99 | | | | | |
| 4–4.99 | | | | | |
| 5–5.99 | | | | | |
| 6–6.99 | | | | | |
| 7–7.99 | | | | | |
| 8–8.99 | | | | | |
| 9–9.99 | | | | | |
| 10–10.99 | | | | | |
| 11–11.99 | | | | | |
| 12–12.99 | | | | | |
| 13–13.99 | | | | | |
| 14–14.99 | | | | | |
| 15–15.99 | | | | | |
| 16–16.99 | | | | | |
| 17–17.99 | | | | | |
| 18–18.99 | | | | | |
| 19–19.99 | | | | | |
| 20–20.99 | | | | | |

Population Simulation #2—Dice Data

Name _____   Section _____

**TABLE 34.7**
Table for Population Simulation #3.

| Population Simulation #3—Dice Data ||||
|---|---|---|---|
| x | S(x) | D(x) | $\ell(x)$ |
| 0–.99 | | | |
| 1–1.99 | | | |
| 2–2.99 | | | |
| 3–3.99 | | | |
| 4–4.99 | | | |
| 5–5.99 | | | |
| 6–6.99 | | | |
| 7–7.99 | | | |
| 8–8.99 | | | |
| 9–9.99 | | | |
| 10–10.99 | | | |
| 11–11.99 | | | |
| 12–12.99 | | | |
| 13–13.99 | | | |
| 14–14.99 | | | |
| 15–15.99 | | | |
| 16–16.99 | | | |
| 17–17.99 | | | |
| 18–18.99 | | | |
| 19–19.99 | | | |
| 20–20.99 | | | |

Name _____  Section _____

## Part V: Survivorship Curves for the United States and Guatemala

Quality of life conditions may impact mortality rates among human populations. Plot the data from the Life Tables for the United States and Guatemala [x versus $\ell(x)$] on graph paper.

*Note:* If we were to compare survivorship rates of females, they would be slightly higher in both countries due to longer life spans seen in women. However, the difference in survivorship between the United States and Guatemala would be similar in males and females.

**TABLE 34.8**
Life Table for the United States and Guatemala

| Life Tables for Male Survivorship | | |
|---|---|---|
| x | United States—$\ell(x)$ | Guatemala—$\ell(x)$ |
| 0 | 1.00 | 1.00 |
| 1 | .973 | .892 |
| 5 | .969 | .805 |
| 10 | .966 | .779 |
| 15 | .964 | .766 |
| 20 | .957 | .748 |
| 25 | .949 | .728 |
| 30 | .940 | .704 |
| 35 | .930 | .677 |
| 40 | .916 | .648 |
| 45 | .895 | .614 |
| 50 | .862 | .572 |
| 55 | .812 | .523 |
| 60 | .739 | .469 |
| 65 | .642 | .392 |
| 70 | .518 | .310 |
| 75 | .385 | .222 |
| 80 | .252 | .149 |
| 85+ | .133 | .085 |

Name _____  Section _____

**Part VI: Discussion**

1. Which type of survivorship curve do the soap bubbles best approximate? Explain the meaning of this curve.

2. Which type of survivorship curve do the dice in Simulation #3 best approximate? Explain the meaning of this curve.

3. After the elimination of AIDS, is there an increase in heart disease? Why or why not?

4. Given that the U.S. is a developed nation and Guatemala is a developing nation, discuss some possible reasons for the differences in the two survivorship curves.

Name _____  Section _____

# LAB 35

# Water Ecology Study

**Problem**

Is there evidence of pollution in water samples collected from the local area?

**Objectives**

After completing this exercise, the student will be able to:

1. Distinguish between physical and biotic factors in the environment.
2. Give an example of a food chain found in a local stream or pond.
3. Identify each trophic level of a food chain as a producer, consumer, or decomposer; give an example of each.
4. Determine if the following are at a concentration high enough to be considered aquatic pollutants:
   a. ammonia
   b. coliform bacteria
   c. chloride
   d. nitrate
   e. nitrite
   f. phosphate.
5. Discuss the importance of adequate levels of oxygen and carbon dioxide.
6. Discuss how pH can effect water organisms.
7. Measure the hardness and the pH of a water sample
8. Describe how microorganisms can effect water quality.

**Preliminary Information**

States bordering on the Great Lakes are fortunate to live by the world's largest supply of fresh water. But concern is growing about increased levels of several pollutants in the drinking water. For example, PCB's or polychlorinated-biphenyl used in industry have entered the food chain. It is recommended that children and pregnant women limit their consumption of fish caught in Lake Michigan to two per year.

Pollution can occur from natural causes, but most water pollution stems from human causes. These range from toxic metals and dyes of industry to silts and acids of mine drainage, from detergents and solid wastes of sewage to potentially dangerous bacteria and chemicals from fertilizers, pesticides and herbicides. One of the results of accelerated pollution is found in the choking of a lake or stream caused by extensive weed and algae growth. This process, called **eutrophication**, is hastened by increased nitrate and phosphate levels brought on by incomplete waste treatment, or fertilizer run-off from lawns and fields.

Every ecosystem has organisms at different feeding levels or **trophic levels**. The **first trophic level** is composed of photosynthetic producers (cyanobacteria, algae, and plants) which make their own food. The **second trophic level** (consumer level I) contains herbivores which feed on the producers. The **third tophic** level (consumer level II) contains **primary carnivores** which feed on trophic level II. The **fourth trophic level** (consumer level III) contains **secondary carnivores** which feed on the third trophic level. **Scavengers** feed on dead organisms from all trophic levels.

These trophic levels can be arranged into a food pyramid with the producers forming the base of the pyramid and each trophic level set in turn on the level below it. Most ecosystems do not have more than four trophic levels because there is not enought energy available to support more than four.

## Part I. Materials and Methods

Samples needed for testing can be collected by the students or made available in the laboratory. Student teams of 4 each, going to a pond or a stream need to assemble the following items:

1. Water quality kits.
2. 4 small jars with lids, with masking tape, label surface water, bottom water, substratum and plankton net.
3. 1 thermometer in a metal case.
4. 1 plankton net (looks like a wind sock with a test tube on the end).
5. 1 bucket (stream only).
6. 1 plastic bag, 1 gallon-size (stream only).
7. 1 pair of waders (stream only).

## Part IA. In the Field at a Pond

1. At the pond, collect the following samples. Keep the lid on the jar, take to the desired depth and remove the lid. Recap and bring to the surface.
    a. 1 sample of the surface water.
    b. 1 sample of the bottom water.
    c. 1 sample of the mud from the water at the bottom of the pond.
    d. 1 sample of water from the plankton net pulled through the surface water. Empty contents of the test tube into a jar.
2. Measure the temperature of the air and the water.
3. Check water quality using test kits.

## Part IB. In the Field at a Stream

1. At a stream, collect the following samples:
    a. 1 sample of surface water at the edge.
    b. 1 sample of bottom water in the middle.
    c. 1 algae-covered rock in bucket, covered with water.
    d. 1 sample of water from the plankton net or near the surface. Empty the contents of the test tube into a jar.
2. Measure the temperature of the air and the water.
3. Check water quality using test kits.

## Part II. In the Lab

1. Collect class data on all chemical tests.
2. Collect class data on all organisms.
3. Place each organism in its proper trophic level.

Name _____   Section _____

## Part III. Physical and Chemical Analysis

Using test kits, several physical factors of the water will be measured. You will be assigned specific kits to use. Record data on the data page and record the data on the board for class use.

### 1. Ammonia (.02-.06 mg/L)

Ammonia is a product of the microbiological decay of animal and plant protein. It can be reused directly to produce plant protein and is used commonly in commercial fertilizers.

Presence of ammonia nitrogen in raw surface water usually indicates domestic pollution. However ammonia nitrogen in ground waters is normally due to natural microbial reduction processes. Its presence in waters used for drinking purposes may require the addition of large amounts of chlorine which will first react with all the ammonia present to form chloramines.

### 2. Carbon Dioxide (10 mg/L or less at surface)

Carbon dioxide occurs naturally in water as a product of aerobic or anaerobic decomposition of organic matter; it also is absorbed readily from the atmosphere. Carbon dioxide reacts with water to form carbonic acid. Although the carbon dioxide concentration usually found in water appears to have no physiological effects on humans it has a marked effect on fish and other aquatic life. Continual exposure to concentrations of 100 mg/L or more has been shown to be fatal to many types of freshwater organisms.

### 3. Chloride (max. 250 mg/L)

Many rocks contain chloride so its presence in the water may be due to natural processes. Heavy salting of the roads during the winter will increase the amount of chloride in the area as can industrial wastes.

### 4. Hardness

Hardness represents the concentrations of magnesium and calcium ions. Water passing through soil and rock will dissolve magnesium and calcium. Soft water contains considerable amounts of chloride and sulfate ions which causes magnesium and calcium ions to precipitate out of solution.

|  | Amount Dissolved Minerals |
|---|---|
| soft water | 0-60 mg/L |
| moderately hard water | 61-120 mg/L |
| hard water | 121-180 mg/L |
| very hard water | over 180 mg/L |

Hardness of water does not indicate pollution.

### 5. Nitrate (10 mg/L)

Nitrate represents the most completely oxidized state of nitrogen commonly found in water. Nitrate-forming bacteria convert nitrites into nitrates under aerobic conditions and lightning converts large amounts of atmospheric nitrogen ($N_2$) directly to nitrates. Many granular commercial fertilizers contain nitrogen in the form of nitrates.

High levels of nitrate in water indicate wastes from run off of heavy fertilized fields. Nitrates can degrade water quality by encouraging excess growth of algae. Drinking waters containing excessive amounts of nitrates can cause infant methemoglobinemia (blue babies). For this reason, a level of 10 mg/L nitrate has been established as the maximum allowable concentration of nitrates in public drinking water supplies.

Natural concentrations rarely exceed 10 mg/L and are often less than 1 mg/L.

## 6. Nitrite (.1 mg/L)

Nitrate nitrogen occurs as an intermediate stage in the biological decomposition of compounds containing organic nitrogen. Nitrite-forming bacteria convert ammonia under aerobic conditions to nitrites. The bacterial reduction of nitrates can also produce nitrites under anaerobic conditions. Nitrites are often used as corrosion inhibitors in industry process water and cooling towers; the food industry uses nitrite compounds as preservatives.

Nitrites are not often found in surface waters where they are readily oxidized to nitrates. The presence of large quantities of nitrites indicates partially decomposed organic wastes in the water being tested. Drinking water concentrations seldom exceed 0.1 mg/L nitrite.

## 7. Oxygen (5-12 mg/L)

The amount of dissolved oxygen present in a given volume of water, that is, the solubility of the oxygen, depends both on temperature and atmospheric pressure. As the temperature increases, the amount of oxygen that water can hold decreases. As the atmospheric pressure increases, the amount of oxygen that water can hold increases.

Because the oxygen saturation of water is dependent on both temperature and atmospheric pressure, the amount of dissolved oxygen required to make a given volume of water 100% saturated also varies. As determined by testing, at 0° C and 760 mm of pressure, a glass of water left undisturbed will contain approximately 14.16 parts per million (ppm) or 14.16 mg/L of dissolved oxygen; this is said to be 100% saturated. Increase in temperature of the water will cause the oxygen level to decline.

Both the shallowness of streams and the constant swirling and churning of water over riffles and falls cause a high degree of contact between the water in streams and the atmosphere. It is not surprising, then, that such streams usually come close to being 100% saturated with oxygen, even when no green plants can be found. It is only in deep holes or polluted water that the amount of dissolved oxygen in streams shows any significant decline.

Under natural conditions the oxygen saturation of streams usually varies little and the organisms that populate the water require considerable oxygen to support life. These organisms cannot tolerate reduced oxygen levels and stream communities are modified dramatically by organic pollution which reduces availability of oxygen. It is this factor of oxygen dependence which makes animals such as trout and mayflies good biological indicators of water quality.

Generally, 5 mg/L dissolved oxygen content is a borderline concentration if considering an extended time period. For adequate game fish population, the dissolved oxygen content should be in the 8-12 mg/L range.

## 8. pH (6.0-9.0)

The pH values below 5 or above 9 are definitely harmful to many animals. In addition, within the normal range, pH can affect the toxicity of poisons. For example, ammonia ($NH_3$) is more toxic in alkaline water than in acidic water, and cyanides and sulphides are more toxic in acidic water than in alkaline.

pH also is thought to be a limiting factor for algae and for some invertebrates. It is known that some crustaceans can withstand a wide range of pH values while others are confined to a narrow range.

The pH of many fresh-water ponds and lakes has changed due to acid rain. Acid rain occurs when nitrogen oxides and sulfur oxides in the air combine with rainwater to form acids such as nitric acid and sulfuric acid. The nitrogen oxides and sulfur oxides are produced in part by the burning of fossil fuel for transportation and electricity.

## 9. Phosphate (.1 mg/L)

Phosphates enter the water supply from biological wastes and residues, agricultural fertilizer run-off, water treatment, industrial effluents, chemical processing, and the use of detergents contribute significantly.

A certain amount of phosphate is essential to organisms in natural waters and often is the limiting nutrient for growth. Too much phosphate can produce eutrophication or overfertilization of receiving waters, especially if large amounts of nitrates are present. The result is the rapid growth of aquatic vegetation in nuisance quantities, and an eventual lowering of the dissolved oxygen content of the lake or stream due to the death and decay of the aquatic vegetation.

## 10. Substratum

Bottom type plays a significant role in determining organisms present in a stream and can determine the dissolved mineral content. Sandy bottoms are the least productive as there is little substrate for either protection or attachment. Bedrock can be non-productive as organisms are totally exposed to the current. Gravel and rubble bottoms are the most productive. There are large areas for attachment sites and rocks provide abundant nooks and crannies where organisms can hide.

## 11. Temperature

Temperature of rivers and streams vary much more rapidly than those of lakes, but this variation is usually over a smaller temperature range than that of still waters.

The temperature of a stream or lake is very important in determining its species composition. Different species have different temperature minimums and maximums that they can tolerate.

Below are some maximum temperatures tolerated by these various fish:

| Fish | Max Temp |
|---|---|
| Carp | 36-37° C |
| Perch | 30° C |
| Pike | 29° C |
| Brook Trout | 25.3° C |
| Rainbow Trout | 24.5° C |

It should be noted that even though a fish can survive at a given temperature, this does not mean that it will have optimal growth or be able to reproduce.

The temperature that fish can withstand depends on such factors as species, age, condition, oxygen content of the water, pH, and the chemical composition of the water.

Temperatures also have a direct affect on the toxicity of poisons to fish. In general, for a given concentration of poison, a rise of 10° C will cut the survival time of fish in half.

Name _____ Section _____

### 12. Bacteria

Bacteria are present all over the earth, so their presence in a water sample would be expected. However, if the concentration of bacteria is very high or specific types of bacteria are present, this is an indication of pollution. Coliform bacteria are naturally present in the digestive tracts of vertebrates. High concentrations of coliform bacteria in a body of water can be caused by such factors as large amounts of fecal matter being put directly into the water, run-off from septic fields during heavy rains, run-off from yards with fecal matter from dogs, run-off from feed lots of livestock or poultry, or run-off from land fills.

### Part IV. Biotic Analysis in the Lab

### 1. Microbial Analysis

A. Obtain two petri dishes, one containing eosin methyl blue agar labelled EMB and the second dish with nutrient agar labelled NA. When introducing the water sample to the petri dish, open the dish a small amount and **close immediately** after adding 2 drops of water. Spread the water over the surface. After 15 minutes, invert the dish and place it in the incubator for 24-48 hours.

B. On the data sheet, sketch the dish with bacterial or fungal colonies. Show the location, size and color of each colony. On the EMB agar, any small, white, glistening colonies could be salmonella which are possible disease-causing contaminants called enteric bacteria. In the U.S., water is considered unfit or contaminated if it contains more than 10 enteric organisms per liter.

Any black or dark green colonies found on EMB agar are lactose fermenting bacteria or *coliform bacteria*, probably *E. coli*. This bacteria is found in the intestine of all mammals. The most abundant mammals in this area are humans. So if the water sample contains this particular bacteria, it is a good indication of sewage pollution.

Sketch the colonies growing on the nutrient agar or NA. Show location, size and color.

### 2. Algae

A. These organisms can range in size from a pin point to huge mats that can cover parts of a pond. Most freshwater algae are green but there are blue-green, yellow or brown organisms. A cell wall generally is visible.

B. Use the keys and diagrams provided to identify each algae type, record the name, trophic level, and level where sample was collected on the data sheet.

C. For algae identification, use the following:
*Pond Life* Golden Guide, pp. 31-38.
*Freshwater Invertebrates,* Needham, pp. 3-10.
*Algae and Water Pollution*, Research Lab, color plates after p. 81.

Name _____   Section _____

### 3. Single-Celled Protist: Protozoa

A. With a pipette, remove a sample from the bottom of the sample jar. Into a petri dish, make many *separate* drops to examine under a binocular scope.
B. Continue doing this sampling until at least one-half of the sample has been examined.
C. For protozoa identification of the single-celled organisms use the following:
   *Pond Life*, Golden Guide, pp. 74-76.
   *Freshwater Invertebrates*, Needham, pp. 13-15.
D. If the specimens are too small to be seen well, switch to a monocular scope and make a wet mount slide of 2 drops. In order to count rapid swimmers, one drop of Lugols solution per 20-40 drops of solution will kill the organisms while leaving them recognizable for counting.
E. Make a sketch of each organism along with its name, type, trophic level and the collection level.

### 4. Larger Organisms

A. Scrape algae off the rock sample into a white enamel pan. Watch for movement and transfer specimen into a petri dish for observations under a binocular scope.
B. For identification, use the following:
   *Pond Life*, Golden Guild, pp. 80-113.
   *Freshwater Invertebrates*, Needham, pp. 18-62.
C. Swirl the liquid in your jar. Put **several separate drops** on the petri dish. Examine under the binocular microscope.
D. Record name, type, trophic level, and collection level of each organism.

Name _____   Section _____

## Pyramid of Trophic Levels in Freshwater Communities

TROPHIC LEVEL I/
PRODUCERS:
*Convert light energy to chemical
energy*

Single-celled Algae
Filamentous Algae
Floating, Submerged, and Emergent
 Vascular Plants

| TROPHIC LEVEL II/<br>CONSUMER LEVEL I:<br>*Feeders on phytoplankton* | TROPHIC LEVEL III/<br>CONSUMER LEVEL II:<br>*Feeders on zooplankton* | TROPHIC LEVEL IV/<br>CONSUMER LEVEL III:<br>*Feeders on larger invertebrates and small fish* |
|---|---|---|
| Protozoa<br>Sponges<br>Rotifers<br>Tardigrades<br>Cladocerans<br>Copepods<br>Haliplid Beetles<br>Mosquito Larva<br>Clams<br>Snails | Planaria<br>Rotifers<br>Nematodes<br>Cladocerans<br>Copepods<br>Blackfly Larva<br>Juvenile Fish | Large Carnivorous Fish such as bass and bluegills<br>Large Frogs<br>Turtles, Snakes<br>Birds<br>Mink |

| *Feeders on larger plants<br>—bryophytes and vascular plants* | *Feeders on larger invertebrates* | SCAVENGERS:<br>*Feeders on dead organic matter, often in particulate form, from all trophic levels* |
|---|---|---|
| Nematodes<br>Crayfish<br>Mayfly Larva<br>Snails<br>Ducks<br>Muskrats | Dysticid Beetles<br>Odonata Nymphs<br>Water Scorpions<br>Dobson Flies<br>Water Bugs<br>Small Fish<br>Frogs, Salamanders<br>Turtles<br>Birds | Protozoa    Cladocera<br>Planaria    Copepods<br>Nematodes  Ostracods<br>Annelids    Amphipods<br>Rotifers    Blackfly Larva<br>Clams |

Name _____   Section _____

## Part V: Data

### Pond Study Data Sheet

Name of Area _____   Team Number _____   Date _____

1. Ammonia _____
2. Carbon Dioxide _____
3. Chloride _____
4. Hardness _____
5. Nitrate _____
6. Nitrite _____
7. Oxygen
8. pH _____
9. Phosphate _____

10. Substratum type _____
11. Temperature °C:

    Water _____

    Air _____

12. Bacteria (number of colonies)

    Nutrient agar _____

    EMB: Salmonella _____

    Coliforms _____

*Record all microbe colonies with a sketch in the space below.

Name _____   Section _____

## Pond Study Data Sheet

Name of Area _____   Date _____

| Name | Type of Plant as "Green Algae" | Trophic Level | Level of Sample |
|------|-------------------------------|---------------|-----------------|
|      |                               |               |                 |

Name _____    Section _____

## Pond Study Data Sheet

Name of Area _____    Date _____

| Name | Type of Animal as "Crustacea" | Trophic Level | Level of Sample |
|------|-------------------------------|---------------|-----------------|
|      |                               |               |                 |

Name _____    Section _____

## Stream Study Data Sheet

Name of Area _____    Team Number _____    Date _____

1. Ammonia _____
2. Carbon Dioxide _____
3. Chloride _____
4. Hardness _____
5. Nitrate _____
6. Nitrite _____
7. Oxygen
8. pH _____
9. Phosphate _____

10. Substratum type _____
11. Temperature °C:

    Water _____

    Air _____

12. Bacteria (number of colonies)

    Nutrient agar _____

    EMB: Salmonella _____

    Coliforms _____

*Record all microbe colonies with a sketch in the space below.

Name _____  Section _____

## Stream Study Data Sheet

Name of Area _____  Date _____

| Name | Type of Plant as "Green Algae" | Trophic Level | Level of Sample |
|------|-------------------------------|---------------|-----------------|
|      |                               |               |                 |

Name _____  Section _____

## Stream Study Data Sheet

Name of Area _____  Date _____

| Name | Type of Animal as "Crustacea" | Trophic Level | Level of Sample |
|------|-------------------------------|---------------|-----------------|
|      |                               |               |                 |

Name _____  Section _____

## Part VI: Discussion

1. Using the data from the physical and chemical analysis and the microscopic examinations of the water samples from a pond, do you consider the pond to be polluted? Why or why not?

2. Using the data from the physical and chemical analysis and the microscopic examination of the water samples from a stream, do you consider the stream to be polluted? Why or why not?

Name _____ Section _____

3. What physical factors of the environment were measured in the lab exercise?

4. What biotic factors of the environment were examined in the lab exercise?

5. What can you conclude from the data obtained by culturing bacteria on two types of agar?

6. What is the relationship between temperature of the water and the amount of dissolved gases in the water?

Name _____  Section _____

7. Our digestive system contains bacteria that don't harm us, so why should the presence of coliform or enteric bacteria in drinking water be of concern to us?

8. In underdeveloped countries, people use water that is inadequate or unsafe. From your survey of factors effecting water pollution, what can be done to make the water safe for humans to use?

9. Did you find any evidence of eutrophication in the samples you studied? Explain. Might these areas test differently at the end of the summer?

10. Why is the pH level of the water important? Why is the pH of many bodies of water in the U.S. and Canada changing?

Name _____ Section _____

Name _____   Section _____

Name _____  Section _____

Name _____  Section _____

Name _____ Section _____

Name _____  Section _____

Biology Practical

1. _____   21. _____

2. _____   22. _____

3. _____   23. _____

4. _____   24. _____

5. _____   25. _____

6. _____   26. _____

7. _____   27. _____

8. _____   28. _____

9. _____   29. _____

10. _____   30. _____

11. _____   31. _____

12. _____   32. _____

13. _____   33. _____

14. _____   34. _____

15. _____   35. _____

16. _____   36. _____

17. _____   37. _____

18. _____   38. _____

19. _____   39. _____

20. _____   40. _____

Name _____ Section _____

Biology Practical

1. _____   21. _____

2. _____   22. _____

3. _____   23. _____

4. _____   24. _____

5. _____   25. _____

6. _____   26. _____

7. _____   27. _____

8. _____   28. _____

9. _____   29. _____

10. _____   30. _____

11. _____   31. _____

12. _____   32. _____

13. _____   33. _____

14. _____   34. _____

15. _____   35. _____

16. _____   36. _____

17. _____   37. _____

18. _____   38. _____

19. _____   39. _____

20. _____   40. _____

Name _____  Section _____

Biology Practical

1. _____  21. _____

2. _____  22. _____

3. _____  23. _____

4. _____  24. _____

5. _____  25. _____

6. _____  26. _____

7. _____  27. _____

8. _____  28. _____

9. _____  29. _____

10. _____  30. _____

11. _____  31. _____

12. _____  32. _____

13. _____  33. _____

14. _____  34. _____

15. _____  35. _____

16. _____  36. _____

17. _____  37. _____

18. _____  38. _____

19. _____  39. _____

20. _____  40. _____

Name _____  Section _____

Biology Practical

1. _____  21. _____

2. _____  22. _____

3. _____  23. _____

4. _____  24. _____

5. _____  25. _____

6. _____  26. _____

7. _____  27. _____

8. _____  28. _____

9. _____  29. _____

10. _____  30. _____

11. _____  31. _____

12. _____  32. _____

13. _____  33. _____

14. _____  34. _____

15. _____  35. _____

16. _____  36. _____

17. _____  37. _____

18. _____  38. _____

19. _____  39. _____

20. _____  40. _____

Name _____  Section _____

Biology Practical

1. _____  21. _____
2. _____  22. _____
3. _____  23. _____
4. _____  24. _____
5. _____  25. _____
6. _____  26. _____
7. _____  27. _____
8. _____  28. _____
9. _____  29. _____
10. _____  30. _____
11. _____  31. _____
12. _____  32. _____
13. _____  33. _____
14. _____  34. _____
15. _____  35. _____
16. _____  36. _____
17. _____  37. _____
18. _____  38. _____
19. _____  39. _____
20. _____  40. _____